KT-379-244

# The CESMM2 Handbook

# The CESMM2 Handbook

A guide to the financial control of contracts using the Civil Engineering Standard Method of Measurement

MARTIN BARNES

Thomas Telford, London, 1986

First published 1986

*Published for the Institution of Civil Engineers by Thomas Telford Ltd, P.O. Box 101, 26–34 Old Street, London EC1P 1JH*

Terms used in this book include terms which are defined in the ICE Conditions of Contract and in the *Civil Engineering Standard Method of Measurement, second edition* (CESMM, second edition). These terms are printed with initial letters in capitals to indicate that the defined meaning is intended. Paragraph numbers and class letters refer to those in the CESMM, second edition; rules in the CESMM, second edition, are referred to by their class and number. The interpretation of the ICE Conditions of Contract and of the *Civil Engineering Standard Method of Measurement*, second edition, offered in this book is not an official interpretation and should not be used as such in the settlement of disputes arising in the course of civil engineering contracts



Barnes, Martin
The CESMM2 handbook: a guide to the financial control of contracts using the Civil Engineering Standard Method of Measurement.
1. Institution of Civil Engineers. Civil engineering standard method of measurement
2. Civil engineering— estimates
I. Title
624'.028'7 TA183

ISBN: 0 7277 0272 6



Typeset by Santype International Ltd, Salisbury, Wilts.
Printed in Great Britain by Redwood Burn Limited, Trowbridge, Wilts.

# *Foreword by the President of the Institution of Civil Engineers*

The *Civil Engineering Standard Method of Measurement* is now well established as an intrinsically important element in contractual procedures, and its usefulness was much enhanced by the guide to the financial control of contracts using the CESMM written for the Institution by Martin Barnes in 1977.

It is entirely appropriate that the publication of the second edition of CESMM should be followed by a second edition of the guide which retains many of the features of the original, but has been extended to take into account the changes in the second edition of CESMM.

Practising Engineers and Surveyors, no less than young engineers and students, will find this compulsive reading, not least because, somewhat rarely, the writer spans both the old and the new and is uniquely able to give guidance on the evolutionary changes that have taken place; I am particularly pleased, as President of the Institution, to commend it not only to young engineers in the early stages of their careers but also to all those engaged in the financial control of contracts.

*D. A. D. Reeve*

# *Acknowledgements*

This book has been given a new title in its second edition. During the life of the first edition, which was called *Measurement in Contract Control*, a large number of people offered me comments and suggestions which have led to changes in the new edition. I am grateful to all of them.

I am particularly grateful to Richard McGill of McGill and Partners and John McGee of Lancashire Polytechnic. They played major roles in helping me to revise the text to accommodate the changes in the second edition of the CESMM itself and also prepared most of the example bill pages.

*Martin Barnes*

Martin Barnes Project Management,
Woodford, Cheshire

*Contents*

# *Introduction*

Financial control means control of money changing hands. Since money almost always changes hands in the opposite direction from that in which goods or services are supplied, it can be considered as the control of who provides what and at what price. This thought establishes a priced bill of quantities as the central vehicle for the financial control of a civil engineering contract. The Bill of Quantities is the agreed statement of the prices which will be paid for work done by the Contractor for the Employer and it shares with the Drawings and the Specification the responsibility for defining what has been agreed shall be done.

Control is usually based on a forecast. The difficulty of controlling something is proportional to the difficulty of predicting its behaviour. The points, finer and coarser, of the financial control of civil engineering contracts revolve around the difficulty the Employer has in forecasting and defining to a Contractor precisely and immutably what he is required to do, and the difficulty the Contractor has in forecasting precisely what the work will cost. To achieve effective control it is necessary to limit these difficulties as much as possible within reasonable limits of practicality. This means using as much precision as possible in defining the work to the Contractor and in enabling him to forecast its cost as precisely as possible. These are the essential functions of bills of quantities. It is the essential function of a method of measurement to define how bills of quantities should be compiled so that they serve these two essential functions.

It is clear from this consideration that a bill of quantities works best if it is a model in words and numbers of the work in a contract. Such a model could be large, intricately detailed and reproducing the workings of the real thing in an exact representation. Alternatively, it could be as simple as possible while still reproducing accurately those aspects of the behaviour of the original which are relevant to the purposes for which the model is constructed.

The first purpose of a bill of quantities is to facilitate the estimating of the cost of work by a contractor when tendering. Considered as a model, it should therefore comprise a list of carefully described parameters on which the cost of the work to be done can be expected to depend. Clearly these parameters should include the quantities of the work to be done in the course of the main construction operations. There is no point in listing those parameters whose influence on the total cost of the work is so small as to be masked by uncertainty in the forecasting of the cost of the major operations.

Other points of general application emerge from this principle of cost-significant parameters. The separation of design from construction in civil engineering contracts and the appointment of contractors on the basis of the lowest tender are the two features of the system which make it essential for a good set of parameters to be passed to contractors for pricing, and for a good set of priced parameters to be passed back to designers and employers. Only then can they design and plan with the benefit of realistic knowledge of how their decisions will affect construction costs. The less contractual pressures cause distortion of the form of the prices exchanged from the form of actual construction costs the better this object is served. It is very much in the interests of employers of the civil engineering industry—whether they are habitually or only occasionally in that role—that the distortion of actual cost parameters should be minimized in priced bills of quantities.

An employer's most important decision is whether to proceed to construction or not. This decision, if it is not to be taken wrongly, must be based on an accurate forecast of contract price. Only if a designer has a means of predicting likely construction cost can such a forecast be achieved. The absence of cost parameters which are sensitive to methods and timing of construction has probably caused as much waste of capital as any other characteristic of the civil engineering industry. It has sustained dependence on the view that quantity is the only cost-significant parameter long after the era when it had some veracity. Generations of contractors, facing drawings first when estimating, have found themselves marvelling at the construction complexity of some concrete shape which has apparently been designed with the object of carrying loads using the minimum volume of concrete. That it has required unnecessary expense in constructing the formwork, in bending and placing the reinforcement and in supporting the member until the concrete is cured often appears to have been ignored.

A major aspect of financial control in civil engineering contracts is the control of the prices paid for work which has been varied. Varying work means varying what the Contractor will be required to do, not varying what has already been actually done. Having once been built, work is seldom varied by demolition and reconstruction; the difficulty of pricing variations arises because what gets built is not what the Contractor originally plans to build. If the work actually built were that shown on the original Drawings and measured in the original Bill of Quantities, the prices given would have to cover all the intricate combinations of costs which produce the total cost the Contractor will actually experience. This would include every hour of every man's paid time—his good days, his bad days and the days when what he does is totally unforeseen. It would include every tonne or cubic metre of material and the unknown number of bricks which get trodden into the ground. It would include every hour of use of every piece of plant—and the weeks when the least popular bulldozer is parked in a far corner of the Site with a track roller missing. The original estimate of the total cost of this varied and unpredictable series of activities could reasonably only be based on an attempt to foresee the level of resources required to finish the job, with many little overestimates balancing many little underestimates. Changes to the work from that originally planned may produce changes to total cost which are unrelated to changed quantities of work. They are less likely to produce changes in cost which are close to the changed valuation if value is taken to be purely proportional to quantity of the finished work. Where there are many variations to the work, the act of faith embodied in the original estimate and tender can be completely undermined. The Contractor may find himself living from day to day doing work the costs of which have no relationship to the pattern originally assumed.

That cost is difficult to predict must not be allowed to obscure the fact that financial control depends on prediction. If the content of the work cannot be predicted the conduct of the work cannot be planned. If the work cannot be planned its cost can only be recorded, not controlled. It must also be accepted that valuation of variations using only unit prices in bills of quantities is an unrealistic exercise for most work and does little to restore the heavily varied project to a climate of effective financial control. Only for the few items of work whose costs are dominated by the cost of a freely available material is the quantity of work a realistic cost parameter. It follows that employers are well served by the civil engineering industry only if contractors are able to plan work effectively: to select and mobilize the plant and labour teams most appropriate to the scale and nature of the expected work and to apply experience and ingenuity to the choice of the most appropriate methods of construction and use of temporary works.

That this type of planning is often invalidated by variations and delays has blunted the incentive of contractors to plan in the interests of economy and profit. The use of over-simplified and unrealistic parameters for pricing variations has led to effort being applied to the pursuit of payment instead of to the pursuit of construction efficiency. In a climate of uncertainty brain power may be better applied to maximizing payment than to minimizing cost.

Mitigation of this problem lies in using better parameters of cost as the basis of prices in bills of quantities. It would be ideal if the items in a bill were a set of parameters of total project cost which the Contractor had priced by forecasting the cost of each and then adding a uniform margin to allow for profit. Then, if parameters such as a quantity of work or a length of time were to change, the application of the new parameters to each of the prices would produce a new total price bearing the same relationship to the original estimated price as the new total cost bore to the original estimated cost. The Employer would then pay for variations at prices which were clearly related to tender prices and the derivation of the adjusted price could be wholly systematic and uncontentious. This ideal is unobtainable, but it is brought closer as bills of quantities are built up from increasingly realistic parameters of actual construction costs.

From the cash flow point of view there are also advantages in sticking to the principle of cost parameters. The closer the relationship between the pattern of the prices in a bill of quantities and the pattern of the construction costs, the closer the amount paid by the Employer to the Contractor each month is to the amount paid by the Contractor each month to his suppliers and sub-contractors. The Contractor's cash balance position is stabilized, only accumulating profit or loss when his operations are costing less or more than was estimated.

Since much of the Contractor's turnover is that of materials suppliers and sub-contractors with little added value, stability and predictability of cash flow has an importance often not appreciated by employers and engineers. Contractors are in business to achieve a return on their resources of management and working capital—a return which is seldom related closely to profit on turnover. Predictability of the amount of working capital required is a function of prompt and cost-related payment from the Employer—another benefit of using pricing parameters closely related to parameters of construction cost.

In the detailed consideration of the financial control of civil engineering contracts and of the use for the purposes of control of the CESMM which follows in this book, the application of the principle set down here is recognized. It should not be thought that the close attention to the affairs of contractors implied by this principle allows them a partisan advantage over employers.

An employer's interest is best served by a contractor who is able to base an accurate estimate on a reliable plan for constructing a clearly defined project, and who is able to carry out the work with a continuing incentive to build efficiently and economically despite the assaults of those unforeseen circumstances which characterize civil engineering work. Confidence in being paid fully, promptly and fairly will lead to the prosperity of efficient contractors and to the demise of those whose success depends more on the vigour with which they pursue doubtful claims.

As Louis XIV's department of works was recommended in 1683, as a result of what may have been the first government enquiry into the financial control of civil engineering contracts: 'In the name of God: re-establish good faith, give the quantities of the work and do not refuse a reasonable extra payment to the contractor who will fulfil his obligations.'

The CESMM sets out to serve the financial control of civil engineering contracts and in doing so finds itself at one with this advice. This book elaborates on and illustrates the use of the CESMM. To facilitate cross-reference its sections correspond to those of the CESMM itself.

The first edition of the CESMM was in use for a few months less than ten years. This book is being published a few months after the second edition of the CESMM was published.

The changes between the first and second editions of the CESMM are many. The layout of what were first called notes has been changed. They are replaced by rules, many of which have the same effect as a former note. The rules are categorized and classified to make reference to them easier and interpretation more straightforward. How they are intended to be used is explained in detail in Section 3.

There is a new class Y in the CESMM second edition which covers sewer renovation work. This is explained in detail in Section 8 as are the changes in those sections of the CESMM which have been heavily overhauled such as class B for ground investigation and class S for rail track.

The principles of the CESMM were not tampered with in producing the second edition. The innovations which CESMM contained such as coded and tabulated measurement rules, simplified itemization and Method-Related Charges were all retained unchanged. Ten years of use of the CESMM had proved their worth. In that ten years, practice in the preparation and use of bills of quantities had developed in the building sector and had, in principle but not in detail, copied all those three innovations in civil engineering practice made with the CESMM.

Many of the changes between the first and second editions of the CESMM are detailed refinements. The purposes behind the more significant of these is explained in the technical sections of this book.

Each section of this book concludes with a schedule of the principal changes between the first and second editions of the CESMM. These schedules do not include the smaller textual changes.

# INTRODUCTION

For brevity, the second edition of the *Civil Engineering Standard Method of Measurement* is referred to throughout this book, as in its title, as CESMM2.

# *Section 1. Definitions*

The definitions given in the CESMM are intended to simplify its text by enabling the defined words and expressions to be used as abbreviations for the full definitions. Where the same words and expressions are used in bills of quantities they are taken to have the same definitions.

There are few changes to the definitions in the second edition of the CESMM. No new definitions are added. Paragraphs 1.12 and 1.13 are altered significantly as described later in this chapter.

It is helpful to use capital initial letters in bills of quantities for any of the terms defined in the CESMM or the Conditions of Contract which themselves have capital initial letters. This leaves no doubt that the defined meaning is intended.

The definition of 'work' in paragraph 1.5 is important. Work in the CESMM is different from the three types of 'Works' defined in clause 1 of the Conditions of Contract. It is a more comprehensive term and includes all the things the Contractor does. It is not confined to the physical matter which he is to construct. For example, maintaining a piece of plant is work as defined in the CESMM, but would not be covered by any of the definitions of Works in the Conditions of Contract.

The expression 'expressly required', defined in paragraph 1.6, is very helpful in the use of bills of quantities. It is used typically in rule M15 of class E which says that the area of supports left in an excavation which is to be measured shall be that area which is 'expressly required' to be left in. The corollary to this rule is that, if the Contractor chooses to leave in supports other than at the Engineer's request, their area will not be measured and the Contractor will receive no specific payment for it. Knowing this, the Contractor will leave the supports in only if it is in his own interests to do so. However, if the Engineer orders the supports to be left in, it is because it is in the Employer's interest that they should be left in, and the Employer will then pay for them at the bill rate for the appropriate items (E 5 7–8 0).

Notes in the CESMM use 'expressly required' wherever it is intended that the Engineer should determine how much work of a particular type is to be paid for. Usually this is work which could be either temporary—done for the convenience of the Contractor—or permanent—done for the benefit of the Employer. Expressly required also means 'shown on the Drawings' and 'described in the Specification' so that it is applicable if intentions are clear before a contract exists.

Rule M5 of Class E is an example which refers to excavation in stages. The Contractor may decide to excavate in stages for his own economy and convenience, or the Engineer may require excavation to be carried out in stages for the benefit of the completed Works. The use of the term 'expressly required' in rule M5 of class E and in other places draws attention to the fact that the work referred to, when done on the Site, will not necessarily be paid for as a special item unless it is a result of work of the type described having been expressly required.

It is prudent for Engineers and Contractors to make sure that work done on the Site which may or may not be measured is carefully recorded between the Contractor and the Engineer's Representative so that it can be agreed whether work is or is not expressly required. Preferably these agreements should be reached before work is carried out.

The definition of 'Bill of Quantities' in paragraph 1.7 establishes that the bill does not determine either the nature or the extent of the work in the Contract. The descriptions of the items only identify work which must be defined elsewhere. This definition is important as it directly implies the difference between the status of the Bill of Quantities under the ICE Conditions of Contract and under the building or JCT conditions of contract.* In the building contract the

* Joint Contracts Tribunal. *Standard form of building contract. Private edition with quantities.* Royal Institute of British Architects, London, 1980.

bill of quantities is the statement of what the Contractor has to do in terms of both definition and quantity. This difference from civil engineering practice is significant. There is no reason to lessen this difference. Both the ICE Conditions of Contract and the CESMM rely heavily on it. The estimator pricing a civil engineering bill of quantities will derive most of the information he needs in order to estimate the cost of the work from the Drawings and Specification. He will use the bill as a source of information about quantities and as the vehicle for offering prices to the Employer. The main vehicles for expressing a design and instructing the Contractor what to build remain the Drawings and Specification.

The definitions of four surfaces in paragraphs 1.10–1.13 are used to avoid ambiguity about the levels from which and to which work such as excavation is measured. When the Contractor first walks on to the Site, the surface he sees is the Original Surface. During the course of the work he may excavate to lower surfaces, leaving the surface in its Final position when everything is finished. Between the Original and Final Surfaces he may do work covered by more than one bill item, as for example in carrying out general excavation before excavating for pile caps. The surface which is left after the work in one bill item is finished is the Excavated Surface for that item, and the Commencing Surface for the work in the next item if there is one. 'Final Surface' is defined as the surface shown on the Drawings at which excavation is to finish. This is so that any further excavation to remove soft spots can be referred to as 'excavation below the Final Surface'. Fig. 1 illustrates the use of the four definitions of surface.

The words of definitions 1.12 and 1.13 were changed in the second edition of the CESMM. No change to the effect of the definitions was intended, but it had become apparent that some compilers of bills of quantities were taking the implied instructions in paragraph 5.21 and the two definitions further than was necessary as regards measurement of excavation of different materials.

It was never intended that the Commencing and Excavated Surfaces of layers of different materials within one hole to be excavated should be identified separately as regards Commencing and Excavated Surfaces. For example, if excavation of a hole involved excavation of a layer of topsoil, then ordinary soft material, then a narrow

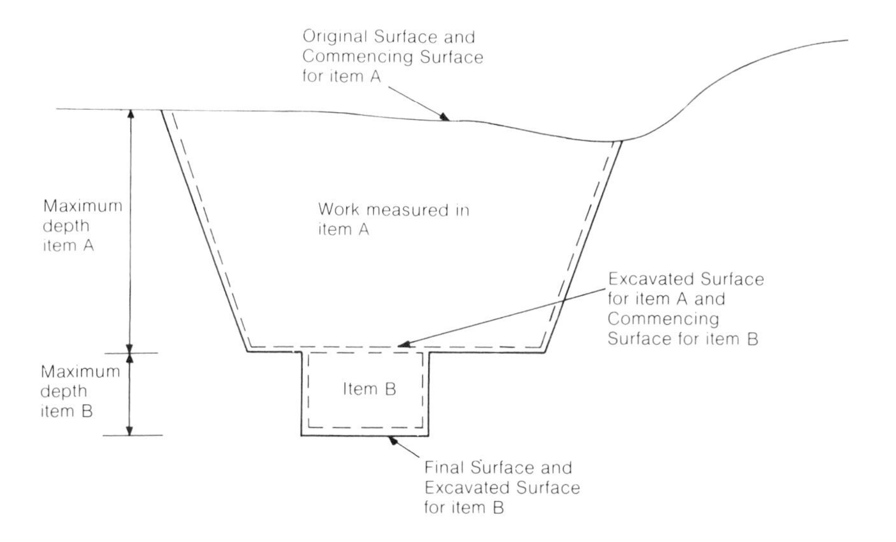


Fig. 1. Application of the definitions of the four surfaces given in paragraphs 1.10–1.13. The Excavated Surface for one item becomes the Commencing Surface for the next item if excavation is measured in more than one stage (see also paragraph 5.21)

band of rock, the Commencing Surface for all three items could be properly regarded as the Original Surface, and the Excavated Surface for all three items could be properly regarded as the Final Surface. The maximum depth stated in the item descriptions for all three items would be the range of those stated in the class E table in which the maximum depth of the complete hole occurred, irrespective of the thickness of each layer of material or of the sequence within the total depth in which they occurred. Fig. 2 illustrates this point.

This produces item descriptions which can occasionally seem peculiar—such as 'Excavate topsoil maximum depth 5–10 m'. This description seems less peculiar as soon as it is understood that it means 'Excavate the topsoil encountered in the course of digging a hole whose maximum depth is between 5 and 10 m'.

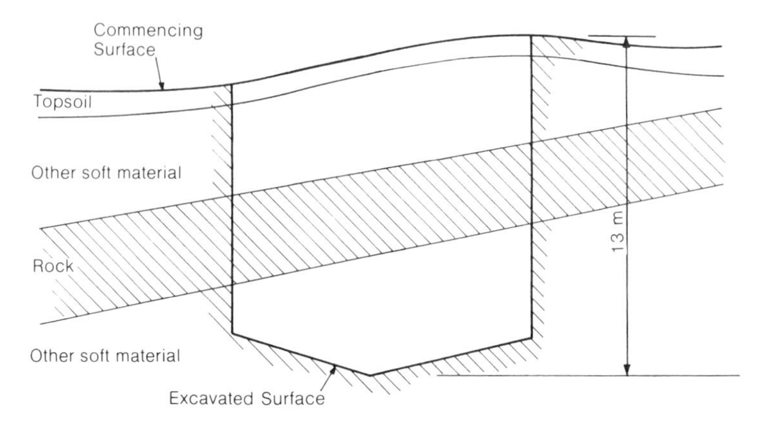


Fig. 2. Three items are required for this excavation. All can be described as 'maximum depth 10–15 m'. Definition rules 1.12 and 1.13 do not require intermediate surfaces to be identified

The changes to definitions 1.12 and 1.13 mean that item descriptions will no longer be cluttered with unnecessary identifications of Commencing and Excavated surfaces. Such phrases as 'Excavate topsoil, Excavated Surface underside of topsoil' or 'Excavated rock, Commencing Surface underside of soft material' should no longer appear. They were never necessary: the new wording of definitions 1.12 and 1.13 makes this more obvious.

Definition 1.14 provides a simple abbreviation for phrases like 'exceeding 5 m but not exceeding 10 m'. In bills compiled using the CESMM this phrase should be abbreviated to '5–10 m'. This convention does not work if ranges are defined with the larger dimension first, e.g. 10–5 m means nothing.

*Schedule of changes in CESMM2*

1. Paragraph 1.1. Makes the definitions apply to terms used in bills of quantities.
2. Paragraph 1.2. Refers to the latest edition of the ICE Conditions of Contract.
3. Paragraphs 1.12 and 1.13. Revised to eliminate the identification of surfaces between different materials in excavation.

# *Section 2. General principles*

The general principles in the CESMM are a small group of rules and statements which set the scene for the detailed rules which follow. Where they are expressed in mandatory terms they are rules of full significance; where they are expressed in less than mandatory terms they give background to help interpretation of the rules.

Paragraph 2.1 points out that the CESMM is intended to be used in conjunction with the fifth edition of the Conditions of Contract and only in connection with works of civil engineering construction. The CESMM can be used with other conditions of contract which invest the Bill of Quantities and the method of measurement with the same functions. The co-ordination of the provisions of the CESMM with other conditions of contract must then be checked carefully before tenders are invited and appropriate amending preamble clauses included in the bill. The standard conditions of contract for ground investigation have been published by the ICE since the first edition of the CESMM was published.* They are referred to in the preface of CESMM2 but not specifically in the general principles. Where the CESMM is used for ground investigation work, the clause numbers can be left as printed because the clauses referred to in the CESMM have the same numbers in both contracts.

In several places the CESMM refers to individual clauses of the Conditions of Contract. Any change to the significance of these references should be checked when the CESMM is used in conjunction with other conditions of contract or with supplementary conditions to the Conditions of Contract. For example, rule C1 of class A of the CESMM limits the coverage of the insurance items included in class

---

* *ICE conditions of contract for ground investigation.* Thomas Telford, London, 1983.

A to the minimum insurance requirements stated in clauses 21 and 23 of the Conditions of Contract. It is therefore essential to use additional specific items to cover any other requirements for insurance which may be applied in particular contracts.

Provided that the responsibilities of the parties are similar and that the status of the Bill of Quantities and method of measurement are similar, the CESMM can be used with other conditions of contract and for work which is not civil engineering. In such cases it will usually be necessary to give amending preambles. There is clearly no point in using the CESMM if the work in a contract is not principally made up of the things which the CESMM covers.

Paragraph 2.2 deals with the problem of identifying and measuring work which is not covered by the CESMM, either because it is work outside the range of work which the CESMM covers or because it is work not sufficiently common to justify its measurement being standardized in the CESMM. No rules are given for itemization, description or measurement of such work but principles are given which should be followed. If the work needs to be measured, that is to say a quantity calculated, any special conventions for so doing which it is intended shall be used should be stated in the Preamble to the bill.

The last sentence of paragraph 2.2 says that non-civil engineering work which has to be covered shall be dealt with in the way which the compiler of the bill chooses, governed only by the need to give the itemization and identification of work in item descriptions in sufficient detail to enable it to be priced adequately.

Paragraph 2.2 does not imply a standard method of measurement because for this type of work there is no necessity for there to be a standard method. Thus, an entry in the Preamble to the bill which complies with this paragraph might refer to another standard method of measurement, such as the standard method of measurement for building,* or it might state a measurement convention adopted for a particular work component. An example of this would be the measurement of large oil tanks associated with oil refinery installations. These are not mentioned in the CESMM but they

* Royal Institution of Chartered Surveyors and National Federation of Building Trades Employers. *Standard method of measurement of building works*, 6th edn. Royal Institution of Chartered Surveyors and National Federation of Building Trades Employers, London, 1978.

might have to be measured within a civil engineering contract. In such a case the compiler of the bill would probably decide to measure the tanks by their mass of steel and might need to state related measurement conventions in the Preamble to the bill. These conventions might include the rules by which the mass of steel in the oil tank was to be calculated for payment.

There is no need for non-standard measurement rules to be complicated or indeed to be given at all in many cases. The function of a bill item is to identify work and to enable a price to be set against it. If, for example, an item description read: 'The thing described in Specification clause 252 and shown in detail D of drawing 137/65' and were given as a sum to be priced by the Contractor it would be a satisfactory item from all points of view. It would not require any measurement conventions. No method of measurement is required for any self-contained component of the work which does not have a particular quantity as a useful parameter of its cost. It might be a plaque on the wall by an entrance, or a complicated piece of manufactured equipment peculiar to the use to be made of the finished project. Only if it is something which may be changed and the financial control of which would benefit from remeasurement of a cost-related quantity is it necessary to give a quantity for it in the bill. Usually these items are something like others which are covered by the CESMM and related measurement rules can be used.

It is good practice to keep away from non-standard measurement conventions for unusual work as much as possible because they can become contentious in preparing final accounts. It is better to make sure that the work required to provide the unusual thing is clearly defined in the Contract and for it to be priced as a sum. If there are several items they can be counted and measured by number. Paragraph 5.18 of the CESMM provides a general convention of measurement, namely that quantities are measured net from the Drawings. Special conventions are needed for non-standard work only when this general convention is inappropriate.

The building work incidental to civil engineering contracts is more difficult to deal with. Compilers of bills are often reluctant to use the building standard method of measurement for an incidental building such as a gatehouse, valve house, or a superstructure in a water or sewerage treatment works. Although use of another standard leads to inconsistency between parts of a bill and some problems of mismatch with the Conditions of Contract, it should be considered when the building work is substantial. When the building work is self-contained and uncomplicated it is controlled most effectively by being treated as specialist work. A provisional sum in the main bill can then be used in a manner appropriate to the circumstances. There should be few problems if it can be priced on a complete and well thought out design which does not have to be changed. Effort is better applied to achieving stable design than to setting up contractual arrangements and drawing up a detailed bill of quantities for something ill-defined which may not be what is eventually wanted.

The scope of the CESMM is referred to in general terms in paragraph 2.2. In detail it can be judged by examination of the various classes of work and the lists of classified components of work which each class includes. It should be noted that the CESMM only provides procedures for measurement of work which is normal new construction. Maintenance and alterations to existing work are not mentioned. Extraction of piles is not mentioned, nor is any other work which involves removal of previous work unless that work is classed as demolition. Any such activity which is to be included in a Contract must be included in the Bill of Quantities. It is suggested that the itemization and description of such work should follow the principles of the appropriate class of the CESMM for new work, with the fact that the items are for extraction or removal stated in descriptions or applicable headings. The new class in CESMM2 dealing with sewer renovation is an exception to the general principle that the CESMM only provides for new construction. It does not indicate that the general principle has been abandoned. There are still no rules for renovation or removal of any other type of existing work. A purist could say that excavation of artificial hard materials, demolition and a few other items came into this category.

When the compiler of a bill is faced with a drawing which shows a component of work not mentioned in the CESMM, he will occasionally be in doubt as to whether it is outside the scope of the CESMM or whether it is within its scope but not mentioned because a separate bill item for it is not required. No standard method of measurement could avoid this dilemma entirely as there will always be some items of work whose nature is just outside the normal understanding of the terms used to name work items in the method. The CESMM should yield few instances of this dilemma, but where it does arise, the treatment should always be to insert non-standard items which describe the work clearly and preferably also to state the location of the work. The resulting bill cannot be held to be in error as this treatment is precisely that intended by paragraphs 2.2 and 5.13.

The objects of the Bill of Quantities are set down in paragraphs 2.4–2.7. Paragraph 2.5 encapsulates the theme of this book. It is important to the financial control of civil engineering contracts that the people who influence it should concern themselves with the practical realities of the costs to contractors of constructing civil engineering Works. Paragraph 2.5 establishes an overriding principle that Bills of Quantities should encourage exposure of cost differentials arising from special circumstances. It is the application of this principle which puts flesh on the standardized skeleton of a bill of quantities and makes the descriptions and prices particular to the job in hand. Its importance is hard to over-emphasize. Differences in construction costs due to the influence of location and other factors on methods of construction are often much greater than the whole cost of the smaller items of permanent work which are itemized separately in a bill.

The items dealing with pipe laying and drain laying provide an example of this. Many bills used to include a schedule of trench depths for drain laying so that an appropriate slightly different rate could be paid for excavating and backfilling trenches whose depth were slightly different, perhaps 100 mm shallower, than that originally billed. In practice the cost of the work is so dependent on whether or not the trench can be battered, whether or not it is through boulders, whether or not there is room to side pile the spoil, whether or not adjacent buildings prevent a backacter swinging round and so on that differences in depth make little impact on the difference between the actual cost and the bill rate.

That cost significance is the all-important factor in dividing the work into separate bill items is demonstrated by how the CESMM deals with this matter. Lengths of pipe laying are given in separate items to indicate different locations by reference to the Drawings (rule A1 of class I). Items are subdivided according to trench depth only within depth ranges.

In applying these detailed rules and the general principle in paragraph 2.5 the bill compiler is required to think about construction costs and to divide up the work so that the likely influence of location on cost is exposed in the bill. He will not be able to read the minds of the contractors who will price his bill, who will not in any case be of one mind as to what is or is not cost significant. Paragraph 2.5 confers on the compiler the obligation to use his best judgement of cost significance—while protecting him from any consequence of his judgement being less than perfect. It achieves this by saying that he 'should' itemize the bill in a way which distinguishes work which 'may' have different cost considerations. If he does not foresee cost differentials perfectly, as he cannot, a contractor cannot claim that the bill is in error and ask for an adjustment to payment.

Paragraph 2.5 ends with an exhortation to strive for brevity and simplicity in bills. Bills of quantities are not works of literature: they are vehicles of technical communication. They should convey information clearly and, in the interests of economy, briefly. An engineer writing a bill of quantities should aim to carry the load of communication safely but with minimum use of resources, in the same way as in his design he aims to carry physical load safely with minimum use of resources.

A general principle not stated in the CESMM is the principle that its use is not mandatory. Whether or not a standard method of measurement is a mandatory document used to be a favourite discussion topic, and each view had strong adherents. When using the ICE Conditions of Contract the method of measurement applicable to each contract is the one the title of which is inserted in the Appendix to the Form of Tender: now normally the CESMM. Such an insertion brings clause 57 of the Conditions of Contract into play, giving a warranty (subject to the condition given in the opening words of the clause) that the Bill of Quantities shall be 'deemed to have been prepared and measurements shall be made according to the procedure set forth' in the CESMM. In order to relate properly to this clause the rules of the CESMM are expressed in authoritative terms; they say, for example, that 'separate items *shall* be given' and 'descriptions *shall* include'.

The Employer or Engineer is free to decide not to use the CESMM, but if he decides that he will use it, he must do what it says that he *shall* unless he expressly shows in the Bill of Quantities that he has done otherwise.

There are a few instances where the CESMM says that details of the procedure 'should' or 'may' be followed. In these instances there is no infringement of the CESMM if the procedure is not followed. These less imperative details of procedure are of two types. One type concerns the procedure for bill layout and arrangement, which has no contractual significance whether followed or not; paragraphs 4.3 and 5.22 are examples. The other type is where the bill compiler is encouraged to use his judgement of likely cost-significant factors to decide on such matters as the subdivision of the bill into parts (paragraph 5.8) or the provision of additional item description (paragraph 5.10). Since the compiler cannot accurately make such judgements without an unattainable foreknowledge of the factors which would actually influence the Contractor's costs, he can only be encouraged to do his best, secure in the knowledge that his not having got it quite right will not entitle the Contractor to have the bill corrected later by an application of clause 55(2). This clause is in the background of all the uses of the words 'may' and 'should' in the CESMM. If, for example, paragraph 5.22 had said 'the work items *shall* be set out in columns ruled as follows . . .' a Contractor could have asked for a bill to be corrected if the rate column on a particular page were only 19 mm wide. Whether or not the Engineer would then decide that this affected 'the value of the work actually carried out' is another matter. It would obviously have been pedantic to have made this a rule instead of a suggestion aimed at helpful standardization.

Paragraph 2.6 is a principle newly stated in CESMM2. It is not a new principle, having been implicit in CESMM from the first. It was thought helpful to state it explicitly in CESMM2.

*Schedule of changes in CESMM2*

1. Newly stated principle at paragraph 2.6.
2. Former paragraph 2.6 is renumbered 2.7.

# *Section 3. Application of the Work Classification*

The Work Classification is the framework and structure of the CESMM. It is the main instrument of the CESMM by which co-ordination of various financial control functions is fostered. It is a basic classification of the work included in civil engineering contracts which can be used for all purposes where it is helpful.

Examples of use of the Work Classification outside the CESMM itself are: as the basis of contractors' allocations in cost control, as an index for records of prices to help with pre-contract estimating and as an index to Specification clauses. The classification is a list of all the commonly occurring components of civil engineering work. It starts with the contractual requirement to provide a performance bond and ends with renovation of sewer manholes.

The Work Classification is not a continuous list. It is divided into blocks of entries which have generic names. First the whole list is divided into 25 classes from 'Class A: General items' to 'Class Y: Sewer renovation'. In between are classes for main operations, like class E for earthworks and class P for piling. Each class is divided into three divisions. The first division is divided into up to eight of the main types of work in the class, the second divides each of these up to eight times, and the process is repeated into the third division. Class I for pipes is illustrated in Fig. 3. The first division classifies pipes by the material of which they are made, the second by the nominal bore of the pipe and the third by the depth below the Commencing Surface at which they are laid.

The entries in the divisions are called 'descriptive features' because, when three are linked together—one drawn from each division between the same pair of horizontal lines—they comprise the description of a component of work in the full list of work in the

## CLASS I: PIPEWORK — PIPES

**Includes:** **Provision, laying and jointing of pipes**
**Excavating and backfilling pipe trenches**
**Excludes:** **Work included in classes J, K, L and Y**

| FIRST DIVISION | | | SECOND DIVISION | | THIRD DIVISION | |
|---|---|---|---|---|---|---|
| 1 | Clay pipes | m | 1 | Nominal bore: not exceeding 200 mm | 1 | Not in trenches |
| 2 | Prestressed concrete pipes | m | 2 | 200–300 mm | 2 | In trenches, depth: not exceeding 1·5 m |
| 3 | Other concrete pipes | m | 3 | 300–600 mm | 3 | 1·5–2 m |
| 4 | Cast or spun iron pipes | m | 4 | 600–900 mm | 4 | 2–2·5 m |
| 5 | Steel pipes | m | 5 | 900–1200 mm | 5 | 2·5–3 m |
| 6 | Plastics pipes | m | 6 | 1200–1500 mm | 6 | 3–3·5 m |
| 7 | Asbestos cement pipes | m | 7 | 1500–1800 mm | 7 | 3·5–4 m |
| 8 | Pitch fibre pipes | m | 8 | exceeding 1800 mm | 8 | exceeding 4 m |

Fig. 3. Classification table for pipes in class I. This is the simplest table in the CESMM and shows clearly how the three divisions of classification combine to produce brief descriptions and code numbers for groups of components of civil engineering works. In this case the brief descriptions are amplified in bills of quantities by more specific information given in accordance with the additional description rules in class I

whole classification. Thus class I can generate 512 components in the list, from 'Clay pipes nominal bore not exceeding 200 mm not in trenches' to 'Pitch fibre pipes nominal bore exceeding 1800 mm in trenches depth exceeding 4 m'.

All components can be identified by their position in the classification list if the numbers of the descriptive features are also linked together. Thus the components mentioned in the previous paragraph are numbers 1 1 1 and 8 8 8 respectively of class I. When a class letter is put in front of a number, a code number which identifies the component is produced. Thus I 2 3 4 is a unique reference number for prestressed concrete pipes of nominal bore 300–600 mm laid in trenches of depth 2–2·5 m.

The Work Classification does not subdivide to the finest level of detail at which distinctions might be needed. It does not, for example, subdivide pipes according to the different types of joints which can be used or different specifications of pipe quality. Also the method of trenching is not classified; neither is the type of terrain being crossed. The reason for this is that the Work Classification and the code numbers which go with it are intended to be adopted as the core of wider ranging classification systems used for other purposes and not necessarily standard among different users. This is the main importance of the Work Classification. Linked as it is to the items in the Bill of Quantities, it makes the use of logical data handling much more worthwhile than hitherto on both sides of the contract. When bill items arising from any Employer or Engineer fit into the same classification, the use of an expansion of that classification by the Contractor as a framework for systematic estimating, cost recording, output recording and valuation becomes possible and justifiable. Also bill preparation with the aid of computers, estimating for the Employer linked to prices and escalation factors stored within the Work Classification are made simpler.

Coded classification of items in the Bill of Quantities is the key to the development of modern data handling arrangements used in civil engineering contracts. It makes possible computer assistance across the whole spectrum of financial control. The person taking off quantities from a drawing could begin a process which was wholly computer aided up to the point where a draft-analysed estimate for consideration by a tenderer's estimator was produced. It is not fanciful to foresee that, if a disc of the Bill of Quantities were to be sent out with the invitation to tender, this could be used by the recipient contractors to produce immediately an analysis of the job with tentative prices calculated from output figures and current unit costs drawn from files held on their own computers. The estimator's job would then be to convert these average prices into prices appropriate to the specific job for which the tender was required.

Clearly, sophistication of that order is not necessary, desirable or practicable in all cases and the usefulness of the Work Classification does not depend on such refinement being attained. The discipline of compiling the CESMM for possible use in that way meant that all procedures related to it could be made simple and logical. This has more immediate benefits to organizations that do not aim to be in the forefront of systems development. For example, the fact that bills compiled using the CESMM list items within the classes of the CESMM means that contractors are able to set up simple arrangements for allocating costs to classes and to find that these match self-contained site activities. The simplest possible form of cost monitoring and comparison with valuations is thereby made utterly straightforward, and can be achieved without requiring a small company to use or obtain the help of specialists to set up a new procedure. The classification provides for contractors' indirect and overhead costs to be allocated to class A. Also two of its eight first division features are left unused to provide room for any such costs which do not appear as prices in the Bill of Quantities and which would not become the subject of Method-Related Charges.

The application of the Work Classification to the preparation of Bills of Quantities is explained in section 3 of the CESMM. Paragraph 3.1 requires that each item description should identify the component of work covered with respect to one feature from each division of the relevant class. The following example is given

> 'Class H (precast concrete) contains three divisions of classification. The first classifies different types of precast concrete units, the second classifies different units by their dimensions, and the third classifies them by their mass. Each item description for precast concrete units shall therefore identify the component of work in terms of the type of unit, its dimensions and mass.'

Paragraph 3.1 does not say that the item descriptions shall use precisely the words which are stated in the Work Classification. Bill compilers are therefore not bound to use the words given in the Work Classification; they should use judgement to produce descriptions which comply with paragraph 3.1 without duplicating information. This is particularly noteworthy where more detail of description is required by a rule in the Work Classification than is given in the tabulated and classified lists themselves. For example, a joint in concrete may be measured which comprises a 'plastics waterstop width 210 mm'. That is an adequate description which does not have to be preceded by a statement that the waterstop is made of 'plastics or rubber' or that it is in the size range '200–300 mm'. Compilers of bills of quantities will need to exercise this type of judgement in many instances, most commonly when a rule requires a particular definition to be given in addition to the general definition provided by the Work Classification table. The rule then overrides the tabulated classification so that the latter merely indicates the appropriate code number for the items concerned. The CESMM2 introduced a rule to make this arrangement more explicit. It is rule 3.10.

Another example will serve to emphasize this important point. It shows that the wording of the Work Classification can be simplified in item descriptions without losing meaning and without infringing the rule in paragraph 3.1. The text 'Unlined V section ditch cross-sectional area 1–1·5 m$^2$' would identify an item clearly as derived from the descriptive features for item K 4 6 5. It is not necessary to add to the description the words from the first division of the classification at K 4 * *: 'French drains, rubble drains, ditches and trenches'. This is because the word 'ditch' appears in the second division descriptive feature and it is irrelevant to point out in the item description that the item is from a division which also includes French and rubble drains.

The lists of different descriptive features given are compiled to show the eight most common types of component in each part of the class. They do not attempt to list all types of component in any class. The digit 9 is to be used for any type of component which is not among the eight listed.

Paragraph 3.2 deals with the question of style in item descriptions.

Its point is that the bill, where it is dealing with Permanent Works, should identify the physical measurable things and not attempt to list all the stages of activity which the Contractor will have to go through to produce them. There are good reasons for this apart from brevity. However careful the bill compiler might be in listing the necessary tasks there will always be at least one more he could have added. The risk of listing tasks inconsistently from one item to another is considerable, and if it occurs the Contractor may subsequently allege that he had not allowed for the thing in the item that was incomplete. It is better and contractually proper to rely on the wording of the Specification, Drawings and Conditions of Contract to establish the overriding assumption that the Contractor knew what he had to do to achieve the defined result and either did or did not allow for it in his price, entirely at his own risk. For example, suppose a bill item were worded: 'Supply, deliver, take into store, place in position, temporarily support, thoroughly clean and cast into in situ concrete mild steel channels as specified and in accordance with the Drawings all to the satisfaction of the Engineer and clear away all rubbish on completion rate to allow for all delays and any necessary cutting of formwork and making good.' This description is much less informative than 'Mild steel channels in pumphouse roof beams as detail E on drawing 137/66.'

The latter description is much less dangerous to financial control of the work on behalf of the Employer than the first, not just because something may have been left out of the first, but because the first invites comparison with other item descriptions. If it is not stated that some other component is to be cleaned, the Contractor may contend that he has not allowed for cleaning, despite the fact that it would be unreasonable for him to expect not to have to.

Phrases like 'properly cleaned', 'to a good finish', and 'well rammed' should be avoided like the plague. They are usually masks for sloppy specification. 'To the satisfaction of the Engineer' is a particularly abhorrent phrase to use in a bill of quantities. The Contractor is under a general obligation to 'complete and maintain the Works in strict accordance with the Contract to the satisfaction of the Engineer' (clause 13(1)). Adding these phrases to a bill item description is misguided on two further counts. First the Specification, not the Bill of Quantities, is the place to define workmanship requirements. Second if the Engineer knows what standard of work will bring him satisfaction he should describe it in the Specification; if he does not know what standard he requires, the Contractor assuredly cannot know and cannot estimate the cost of reaching it.

In the context of paragraphs 3.2 and 3.3, it should be noted that the general case, assumed unless otherwise stated, is that item descriptions identify new work which is to be constructed by the Contractor using materials which he has obtained at his cost. Additional description is needed wherever this assumption is not intended. For example, additional description is needed to identify items for extracting piles, for underpinning and related work to existing structures, and for any work which involves the use of existing materials, such as the relocation of existing street furniture.

The first example in paragraph 3.3 uses the expression 'excluding supply and delivery to the Site' to illustrate an item description for work which is specifically limited. Compilers of bills of quantities should note that this expression is not defined in the CESMM as its exact scope may vary from one contract to another. For example, pipework supplied by the Employer may be for collection by the Contractor from a depot or may be delivered to the site for the Contractor. Compilers of bills of quantities should therefore ensure that, whenever the scope of an item is specifically limited in this way, the precise limitation is stated and that, if an abbreviated expression such as 'fix only' is used, a definition of this term is given in the preambles to bills or in a heading to the appropriate items.

In CESMM2 a second example is included in paragraph 3.3. This is intended to clarify how work divided between two bill items should be described. This often occurs when additional description is necessary to make it clear how the cost of supply and fix are divided between two items. The principle to be followed in bills prepared using the CESMM is simple. Unless otherwise stated, all items are assumed to include both supply and fixing of the work they cover. Whenever this is not intended, additional description must be given to identify whether the item is intended to cover supply or to cover fixing.

Paragraphs 3.4 and 3.9 which deal with separate items, are important as they directly govern against which different components of work the tenderer will be able to insert different prices. There is no absolute criterion of 'full' measurement—no level of detailed sub-division of the work into items which is complete or incomplete. A bill of quantities for a motorway would earn the name if it contained one item

| Number | Item description | Unit | Quantity | Rate | Amount | |
|---|---|---|---|---|---|---|
| | | | | | £ | p |
| 1 | Motorway | m | 23124 | | | |

It would also earn the name if it contained separate items for each piece of reinforcing steel of a different shape in each different bridge, for each differently shaped formwork surface, for each detail of water stops and drainage fittings, and generally for no two things of any detectable dissimilarity. Such a bill would contain several thousand items and would be as useless due to its over-complexity as the single item bill would be due to its over-simplicity.

In practice effective financial control is served by using bills of quantities which balance the opposing pressures for precision and simplicity. Real cost differentials should be exposed by dividing work into separate items which it is helpful to price differently. Trivial or imagined cost differentials are ignored when their influence on the amounts of money changing hands does not justify the cost of cosseting the necessary items through the processes of estimating and interim measurement into the final account.

The effect of paragraph 3.4 is that no two items from the Work Classification list may be put into one item. Hence formwork may not be included with concrete, lined and unlined ditches may not be covered in the same item, and so on. Thus the Work Classification, coupled with paragraph 3.4, has the effect which is achieved in other methods of measurement by many different rules of the general form 'separate items shall be given for lined ditches' or for any other distinctive component of work.

Paragraphs 3.6–3.11 are the rules which establish the function of all the material which appears on the right-hand pages of the Work Classification. The changed arrangement of the right-hand pages is the most striking new feature of the second edition of the CESMM. The statements in the first edition which appeared on these pages but which were called notes have been extensively revised in the second edition. They have also been categorized according to whether they refer to measurement itself, definitions of terms used in the Work Classification coverage to be assumed for particular items and description to be given in addition to that which is derived from the main classification tables in accordance with paragraph 3.1. These four types of rules are set out in columns. Generally, each rule is printed alongside the section of the classification tables on the left-hand page to which it refers. Figure 4 is a reproduction of the first group of right-hand page rules from class C of CESMM2.

Paragraph 3.6 is the definition of the first type of rule, the measurement rule. Measurement rules either say something which affects how a quantity against a particular item or group of items will be calculated or say something about the circumstances in which particular work will or will not be measured. The measurement rules exemplified in Fig. 4 show these different functions. Rules M1 and M2 are rules affecting the quantity calculated for such items as the depths of holes in carrying out geotechnical processes. Rule M3 is a measurement rule stating the circumstances in which work associated with drilling grout holes shall be measured. M4 is an example of a measurement rule using the expression 'expressly required'. In CESMM2, all uses of this expression in the Work Classification have become measurement rules.

Paragraph 3.6 refers to paragraph 5.18 and vice versa. The logical connection between the two demonstrates one of the advantages which CESMM2 has over the first edition. Setting out the measurement rules in the new way means that they become quite clearly and consistently the exceptions to the general rule of calculation of quantities set out in paragraph 5.18. Put another way, this means that if there is no measurement rule alongside a particular group of items in the Work Classification the general rule of measurement in paragraph 5.18 applies. This is the principle, but it does not apply entirely. It has to be qualified because there may be measurement rules that are applicable to the whole of one class which appear at the head of the class as illustrated by Fig. 4.

Paragraph 3.7 establishes the function of definition rules. The

**CLASS C**

| MEASUREMENT RULES | DEFINITION RULES | COVERAGE RULES | ADDITIONAL DESCRIPTION RULES |
|---|---|---|---|
| **M1** The Commencing Surface adopted in the preparation of the Bill of Quantities shall be adopted for the measurement of the completed work.<br><br>**M2** The depths of *grout holes*, holes for *ground anchorages* and *drains* shall be measured along the holes irrespective of inclination. | **D1** Drilling and excavation for work in this class shall be deemed to be in *material other than rock or artificial hard material* unless otherwise stated in item descriptions. | **C1** Items for work in this class shall be deemed to include disposal of excavated material and removal of existing services. | |
| **M3** Drilling through previously grouted holes in the course of stage grouting shall not be measured. Where holes are expressly required to be extended, the number of holes shall be measured and drilling through previously grouted holes shall be measured as *drilling through rock or artificial hard material*.<br><br>**M4** The *number of stages* measured shall be the total number of grouting stages expressly required. | | | **A1** The diameters of holes shall be stated in item descriptions for *drilling* and *driving for grout holes* and *grout holes*. |

Fig. 4. The layout of the classified rules in CESMM2: note the different style of each of the four types of rule, the horizontal alignment and the use of the double horizontal line to separate rules of general application to the class.

phrases or words for which definitions are given in the definition rules are assumed to have the same meanings when they are used in bills of quantities. Most of the definition rules cover matters which it is helpful to define in order to avoid ambiguity in bills of quantities. For example, definition rule D7 in class F says that a wall less than 1 m long is to be called a column. It is followed by definition rule D8 which says what is meant by a 'special beam section'. A particular function of some definition rules is to enable bill item descriptions to be abbreviated. The example which appears in Fig. 4 is of this type. It says that drilling and excavation for geotechnical processes shall be deemed to be in material other than rock or artificial hard material unless otherwise stated in item descriptions. The effect of this rule is nothing more than to permit the words 'material other than rock or artificial hard material' to be omitted from bill item descriptions. This mechanism is used in a number of places in CESMM2 to allow the words which would establish the general case to be omitted from bill item descriptions leaving only special cases to be explicitly mentioned.

Paragraph 3.8 is an important new rule in CESMM2 which establishes the function of coverage rules. These replace the notes which appeared in the first edition of the CESMM which said that separate items were not required for particular items of work. An example was note 6 of class R which said that separate items were not required for formwork to concrete pavements. Such notes had the effect, for example, that items for concrete pavements included formwork. In CESMM2, the same point is made by a coverage rule. Coverage rule C1 in class R of CESMM2 says that items in the class which involve in situ concrete shall be deemed to include formwork. The effect is the same, but the new coverage rules are a positive statement which is less ambiguous than the negative statement which appeared in the first edition.

Paragraph 3.8 includes a very important provision to the effect that coverage rules do not state all the work covered by the item concerned. This means, taking the same example again, that items in bills of quantities for concrete carriageway slabs do cover, and the rates entered against such items in a Bill of Quantities are deemd to include, formwork. Paragraph 3.8 is worded carefully so that the existence of this coverage rule cannot lead to an argument that any work other than formwork is not included in the item because it is not mentioned in the rule. It follows that coverage rules only draw attention to particular elements of cost within a bill item which are certainly deemed to be covered. They leave the majority of the elements of cost to be inferred from the description used in the Bill of Quantities which identifies the work shown on the Drawings and described in the Specification. The coverage rule does not, of course, override drawings and specifications in the sense that if, for example, no formwork is required for a particular carriageway slab the rate against a bill item for it would not include formwork. Coverage rules do not normally mention work which is also mentioned in additional description rules. Occasionally this repetition is made in order to add emphasis.

The last sentence of paragraph 3.8 points out that the Contractor may have allowed for the work referred to in a coverage rule in a Method-Related Charge. This is quite permissible and indeed is to be encouraged where the cost of the work referred to in the coverage rule is either independent of the quantity required or related to time.

Paragraph 3.9 in CESMM2 is the former paragraph 3.2 from the first edition combined with the former paragraph 3.6. It establishes the function of the additional description rules in the Work Classification.

The importance of the additional description rules should be emphasized. During the life of the first edition of the CESMM, the significance of the notes which required additional description was often underestimated even after the document had been in use for several years. Paragraph 3.9 is very explicit, and it is to be hoped that further misunderstandings will not happen. In simple terms, it should be understood that the classification tables on the left-hand page in the CESMM only generate the basic subdivision of civil engineering work into items and the basic descriptions which will be used. Further description and further subdivision into items is very often required as a result of applying the additional description rules. The main classification tables are divided into three divisions. It may be helpful to think of the additional description rules as providing a fourth uncoded division.

Paragraph 3.10 is an addition to the CESMM2. It only makes explicit what was implicit before: that additional description rules can override the main classification table. It refers in particular to dimensions mentioned in item descriptions. There are a number of instances in CESMM2, as in the first edition, where an additional description rule requires a particular dimension to be stated in an item description although the related part of the Work Classification table only requires a range of dimensions to be stated. The most well-known example of this is the one which is now used as an example in CESMM2.

> Additional description rule A2 of class I requires that the nominal bores of the pipes shall be stated in item descriptions. The range of nominal bore taken from the second division of the classification of class I shall not also be stated.

Paragraph 3.11 is a new rule illustrated here by Fig. 4. To reduce repetition of the rules in CESMM, any rules which apply to the whole of a class are printed at the head of the first right-hand page above a double horizontal line. If the class runs over onto a second page, a reminder that there are rules of general application is printed at the head of the following right-hand page. In some classes, rules are repeated once or twice within the class because they apply to more than one section of the table. In such cases the rule is printed more than once but the number is kept the same.

The phrases printed in italics in the rules on the right-hand pages of the Work Classification are those which are taken directly from the classification table on the left-hand page. This has no contractual significance: it is adopted merely as a convenience to enable users of the CESMM to recognize very quickly the type of work to which the rule refers.

The numbering of the rules in CESMM2 is changed to reflect the categorization. Rules of each type are numbered consecutively within each class with a prefix letter identifying the type as follows

M Measurement rule
D Definition rule
C Coverage rule
A Additional description rule

Some classes include a note printed below the categorized rules. These are genuine notes in the sense that they refer to options open to compilers of bills of quantities; they are not rules which must be followed. There are notes in classes A, G, K, N, R, T, V and W.

In accordance with the general principles, the main object of dividing civil engineering work into items is to derive a set of items which most realistically represents those different aspects of what is to be done which influence the total cost of what is to be done. This pursuit of realism is tempered only by the need to limit the number of items to one which will not generate clerical work out of proportion to the resulting precision of the estimating for and valuation of the work.

The introduction of Method-Related Charges takes one significant pressure off this process of dividing work into separate items. Since Method-Related Charges are provided to cover, along with the Engineer's general items, all the elements of the cost of the work which are not directly proportional to the quantities of the Permanent Works, it follows that the classification of the Permanent Works need concern itself only with distinctions of those costs which are proportional to the quantities of measured work. This is why, for example, there are no items in the Work Classification for bringing items of plant to and from the Site. Where the cost of such transport of plant is significant, the tenderer may, and should, enter it as a Method-Related Charge.

An important point about the Work Classification is that the Contractor is assumed to have allowed for everything which is required by the Contract somewhere in his set of prices against items, subject only to the provision for correction of errors and omissions from the bill provided by clause 55(2). Neither the CESMM nor the Bill of Quantities attempts to tell the Contractor precisely how he shall distribute his expected costs between the prices inserted against the various bill items. There is only a general assumption that the costs of the components of the work identified by the item descriptions, which are related in some proportion to the measurement unit of the item, are in the prices inserted against each item.

Paragraph 3.5 is the rule which gives effect to the way in which the Work Classification prescribes the units in which the various work components are to be measured.

*Schedule of changes in CESMM2*

1. Work split between two classes is referred to in paragraph 3.3.
2. Rules for the use of the right-hand pages in the Work Classification have been changed and now appear as paragraphs 3.6–3.11.

# *Section 4. Coding and numbering of items*

Section 4 of the CESMM explains how the coding of items in the Work Classification can be used as the basis of item numbering in bills of quantities. It is not a requirement of the CESMM that code numbers should be used as item numbers. It is clearly helpful to contractors using the CESMM classification structure in connection with estimating or cost control if bills arrive ready coded by means of the item numbers. The other advantage of using item numbers based on code numbers is that it encourages uniformity of sequence of items in bills.

Items should not be listed rigidly in ascending order of code number as there are some places where this would prevent useful headings being provided, but it is of great assistance to estimators, planners and buyers if bills arriving from various sources all have items presented in the same sequence. It can take a long time for an estimator or for the buyer responsible for sending out enquiries to develop the necessary familiarity with the document if it is not set out in the order which has become familiar through standardization.

There is a tenacious view held by many who are not directly concerned that estimators like to see work billed in the order in which it is done on the Site, to see the formwork with the concrete or to see the large value items listed first. Most of these ideas are based on a mental image of the estimate being compiled by one person who works through the bill from page one to the end, considering each item separately and writing in its price before reading the next. This image is quite out of date and was probably ever near the truth only in the building sector. Estimating generally begins with separating the materials and sub-contracted elements for obtaining quotations. Then the various operations of plant and labour are planned, information being taken from the Drawings, Specification and Bill of Quantities. Unit estimated costs emerge first, in blocks of items which may be for any part of the bill. Gaps are gradually filled and it is only at a very late stage in the process that any prices are fixed based on these estimated costs. The sequence of items is largely unimportant, but it is of significant convenience if it is always the same.

Bills are usually divided into parts to distinguish phases of the work or different structures, or for other cost-significant purposes. Where separate 'bills' were formerly the normal way of dividing up a bill, these are now replaced by 'parts'. The items in each part are arranged in the general order of the Work Classification. This means that there can be items with the same code number in different parts of the bill. Where items are referred to outside the bill their number is preceded by the number of the part from which they are taken so that no two items in one bill have the same complete reference number. Thus, if an item H 1 3 6 were in two parts of a motorway bill, one of which was 'Part 4: Corvedale Road Bridge', outside the bill the item number would be referred to as 4. H 1 3 6. If that item were to have a suffix number as well, denoting the presence of additional description, the item number would be 4. H 1 3 6.2. In a very big bill there may be many parts and many suffixes and an item number could be 24. H 1 3 6.15. The part of the item number between the dots has the standard meaning drawn from the CESMM, in this case 'Precast concrete beam length 7–10 m mass 5–10 t'. The prefix and suffix are non-standard and are peculiar to the particular bill. The prefix is a part reference, usually to a location, which might mean 'Corvedale Road Bridge' in one job or 'Whettleton pumping station' in another. The suffix number might mean the additional description 'rectangular 200 mm × 400 mm concrete grade 30 mark 25 in deck' in the first case or 'I section 250 mm × 350 mm concrete grade 25 mark 4 in roof' in the second.

Where subheadings are used within parts of a bill it is sometimes difficult to ensure that the same item number is not repeated within a part. This can be avoided by adding suffix numbers as if the heading were additional description. The suffix number 1 may be added to all items which have additional description, whether or not they are followed by other items of the same code with different additional description. This then acts merely to indicate that there is additional description amplifying the standardized description derived from the Work Classification. The CESMM does not require this indication to be given.

Some parts of the Work Classification do not make use of all three divisions. An item drawn from such a part may be coded with zero in the position of the unused division. Examples of this are given in the example bill pages in section 8 of this book. To comply with paragraphs 4.5 and 4.6, the numbers of the gaps in the Work Classification should not be used to code non-classified items. However, this practice cannot mislead users and it is sometimes adopted in order to simplify item numbers.

It is pertinent to end this section with a reminder and a reassurance. The reminder is that the whole subject of code numbers and item numbers has no contractual significance. The only item description is the text in the column headed 'Item description' and it is unaffected by the item number be it correctly coded, incorrectly coded or not coded at all. The reassurance is that estimators are not expected to commit code numbers to memory or to use the CESMM as a code directory. The CESMM gives no authority for code numbers to be used within item descriptions in place of text properly assembled in accordance with the rules of the CESMM.

Section 4 in CESMM2 is the same as in the first edition. Despite the novelty of the coding system introduced in the first edition, no change to how it works was necessary in the second edition. The code numbers themselves have been changed in the Work Classification where the itemization has been changed, but the arrangements for using the coding are unaltered.

*Schedule of changes in CESMM2*

None.

# *Section 5. Preparation of the Bill of Quantities*

Section 5 of the CESMM contains the details of the general rules for preparing bills of quantities and deals with the treatment of special features of bills such as items for Daywork and Nominated Sub-contractors.

The first paragraph is a reminder that the method of measurement is a method of measuring what has been done after it has been built as well as a method of measuring what is expected to be required. The CESMM contains rules for compiling bills of quantities from the measurements made of proposed work. The same rules apply to the measurements made of completed work.

Paragraph 5.2 establishes that bills of quantities should be divided into sections in a standardized way in order to sustain uniformity of presentation which is one of the principal aims of the CESMM. This rule produces the typical table of contents for a sample Bill of Quantities shown in Fig. 5. The sections are identified by the letters A to E so that they are not confused with the locational or cost-significant parts into which the work items are divided and which are identified by numbers. Section C is the Daywork Schedule, and may be omitted.

The use of the list of principal quantities is defined in paragraph 5.3. This paragraph echoes the wording of the Form of Tender so that it is quite clear that no subsequent contractual contentions can be attached to any discrepancy between the impression created by the list of principal quantities and the details of the quantities proper given in the Bill of Quantities itself.

A list of principal quantities like that required by the CESMM was formerly often given in the Specification. There is no standard method of specification, but bringing the requirement into the

Section A. List of principal quantities
Section B. Preamble
Section C. Daywork Schedule
Section D. Work items
  Part 1. General items
  Part 2. Outfall
  Part 3. Plowden treatment works
  Part 4. West branch sewers
  Part 5. East branch sewers
  Part 6. Whettleton pumping station
Section E. Grand Summary

Fig. 5. Example of the standardized sequence of contents of a bill of quantities which results from application of paragraphs 5.2 and 5.8. Sections of the Bill are identified by the letters A to E to distinguish them from the numbered parts into which the work items themselves are divided

CESMM ensures that the list is always given. It does not need to be an extensive list. It can be assembled by the bill compiler very easily after the bill has been drafted, using his knowledge of the job and of the bill layout. It will be referred to by estimators at the beginning of their involvement with the bill, until they also gain a knowledge of the job and of the bill layout.

The CESMM gives no guidance on which quantities should be regarded as principal ones for the purpose of compiling the list because it would be difficult to make a bad job of selecting them. It would be exceptional for the list to need a second page in the bill. If there are too many principal quantities the object of the list is defeated. A suggested list for a hypothetical contract is shown in Fig. 6. It is not essential to divide the list into the bill parts and it is unhelpful to do so if there are very many parts. More detail may be given when one type of construction predominates.

Only four of the functions of the Preamble to a bill of quantities are defined in the CESMM. It is the place where any special methods of measurement adopted for the particular contract are defined. That the CESMM sets forth a procedure to be used where the standard procedure is not going to be used is something of a paradox. It should not be taken as a licence to use non-standard

Section A. List of principal quantities

| | | |
|---|---|---|
| Part 1. General items | | |
| | Provisional Sums | 40 000 £ |
| | Prime Cost Items | 165 000 £ |
| Part 2. Outfall | | |
| | Excavation | 4 000 m$^3$ |
| | Filling | 1 000 m$^3$ |
| | Concrete | 850 m$^3$ |
| | 900 mm steel pipework | 200 m |
| Part 3. Plowden treatment works | | |
| | Excavation | 23 000 m$^3$ |
| | Filling | 8 000 m$^3$ |
| | Concrete | 9 000 m$^3$ |
| | Pipework | 1 200 m |
| Part 4. West branch sewers | | |
| | Pipelines | 8 000 m |
| | Manholes | 40 nr |
| Part 5. East branch sewers | | |
| | Pipelines | 9 000 m |
| | Manholes | 50 nr |
| Part 6. Whettleton pumping station | | |
| | Excavation | 1 200 m$^3$ |
| | Filling | 400 m$^3$ |
| | Concrete | 600 m$^3$ |

Fig. 6. Example of a list of principal quantities compiled to comply with paragraph 5.3

methods of measurement indiscriminately. The stipulated procedure should be used throughout every bill of quantities except where there are practical reasons why a clearly defined different procedure would be preferable.

An example of such a reason is where the Contractor is to have some abnormal design responsibility so that the quantities of work to be carried out will be more under his control than usual. If the Contractor has a performance specification for a component it is preferable for it to be measured in less detail than the CESMM requires so that the valuation of the work is unaffected by the Contractor's choice of design. A similar situation arises where the Specification contains permitted alternatives, as is common for major road contracts. The permitted alternatives have to be grouped into one bill item so that the valuation is unaffected by the choice between alternatives eventually made by the Contractor. Such groupings of work within one item conflict with paragraph 3.4. Where this arrangement is used a clause should be given in the Preamble which states which components of the work are the subject of permitted alternative specifications and which provisions of the CESMM Work Classification are not to apply, and sets out any non-standard rules for measurement which are to apply. A change in CESMM2 is the addition to clause 5.4 of a statement about measurement of contractor-designed work or work where the contractor is expected to make a design decision by the choice between alternative materials or methods. This is another example where CESMM2 makes something explicit which was implicit before. The principle set out is simply that, if the Contractor is given a choice of permanent works to provide, the Contract Sum should be unaltered whatever choice he makes. This implies a briefer than usual description and itemization in the Bill of Quantities to which reference must be given in the Preamble.

The standard method of measurement used before the CESMM was published listed a group of directions to tenderers which were usually repeated in bills. All of these are now covered by statements in the Conditions of Contract, the Appendix to the Form of Tender, or the CESMM and there is no need for them to be given in bills of quantities. It used to be particularly common for the Preamble to say that the cost of any work for which the items were unpriced would be deemed to be covered by the prices inserted against the items which were priced. This point is now established in clauses 11(2) and 55(2) of the Conditions of Contract.

The former practice of stating in the Preamble that the bill had been prepared 'generally in accordance' with the standard method of measurement is something which by now should have died out. It was intended to shield the Employer from any inadvertent deficiencies in the preparation of the bill—by stating that none of the particular rules of measurement could be assumed to have been followed. Such statements were always unhelpful; they are now quite out of place. It is no more tolerable for the Contractor to accept such a declaration than it is for the Employer to receive a tender declaring that the prices are to cover work 'generally in accordance with the Specification'.

The Preamble may also be used to bring into the Contract precise definitions of the tasks which are intended to be covered by the items. Such schedules of item coverage are now well established in some of the sectors of civil engineering where an employer is able to use a standard specification linked to the Bills of Quantities for his work. Item coverage schedules can be brought into bills compiled using the CESMM by setting out in the Preamble the amendment to or substitution for paragraph 5.11 of the CESMM which is to apply. For example, if an existing schedule of item coverage is to apply, it could be stated in the Preamble that paragraph 5.11 of the CESMM is to be amended by the addition at the end of the paragraph of the words 'and the Schedule of Item Coverage issued by . . . dated . . .'

It must be noted that the use of schedules of item coverage is not essential to the use of the CESMM. It is justified only where a large volume of work is being carried out using a standard specification which covers a limited range of work. Where this is not the case item coverage schedules can be misleading. Suppose, for example, a schedule stated that the work covered by the items for kerbs included construction of a concrete bed and backing, including formwork, reinforcement and joints. Provided that the item coverage was used only where a standard detail of beds and backings was shown on the Drawings and described in the Specification this would be a helpful reminder to the estimator. It does not help actually to price the item because the estimator in any case has to look at the Drawings and Specification in order to find out the dimensions and quantities of work and to estimate its cost.

Where there are no standard details or standard specification there is a danger that the standardized item coverage may conflict with the particular Drawings and Specification. It would be most unhelpful for the standard item coverage to say that the item included concrete bed and backing to kerbs if no such work was actually required.

Users of CESMM2 must understand that the renaming of some of the former notes as coverage rules does not mean that the CESMM now includes a full coverage schedule. It only includes a relatively small number of coverage statements made to avoid uncertainty in particular areas.

The Preamble is also the place to include statements tailoring the CESMM to special circumstances. When the work is not to be carried out in the United Kingdom or is to be advertised for tenders internationally, it is helpful to add a clause to the Preamble relating to British Standard specifications. The CESMM refers in places to materials, tests and components which comply with BS specifications. When similar components which are not to BS specifications are required the bill compiler has the choice of either using non-standard item descriptions or of making a general statement in the Preamble to the effect that components complying with equivalent specifications will be accepted.

Paragraph 5.4 no longer requires amendments to the CESMM to be mentioned in item descriptions (or headings) as well as in the Preamble itself. Estimators must now study the Preamble carefully as they cannot rely on a reminder appearing in the bill itself that something is being treated in a non-standard way. In compensation for this change, paragraph 5.4 now requires the extent of the work affected by non-standard measurement to be stated in the Preamble.

Paragraph 5.5 refers to the definition of rock given in the Preamble of the bill for any contract which includes excavation, boring or driving. Excavation here does not mean only work described as excavation in the Work Classification; it includes work which comes within the normal wider meaning of the word, such as trenching for pipes and ducts. The paragraph effectively means that a definition of rock must be given in all but the rare cases where the Permanent Works are above ground superimposed on existing foundations or structures.

Standard methods of measurement used to require the definition of rock to be in terms of the expected geological formations. The CESMM only requires a definition to be given; it does not constrain its terms. Until numerical classifications of the digability of strata can be given, it is advisable that this definition should be in terms of geological formations and conditions. These should be related to the terms used in any borehole logs or other site investigation data made available to tenderers so that the quantities given can be referred to such data. It is not helpful to define rock in terms of the plant which the Engineer considers is capable of removing the various strata shown on the borehole logs unless his assumptions in this respect are stated in the Contract. Boulders should be referred to in the definition of rock. The minimum size of boulder which is classed as rock is stated in class E, but if a similar minimum is intended to be applied to excavation in other classes it should be stated in the definition of rock in the Preamble.

CESMM2 adds two more functions for the Preamble. Paragraphs 6.4 and 7.7 now require that statements about the Adjustment Item and Method-Related Charges should appear in the Preamble. The following statements can be used.

> *Adjustment Item*
> For the purposes of clause 60 interim additions or deductions on account of the amount, if any, of the Adjustment Item shall be made in instalments in interim certificates in the proportion that the amount referred to in clause 60(2)(a) bears to the total of the Bill of Quantities before the addition or deduction of the Adjustment Item.
>
> *Method-Related charges*
> Method-Related Charges shall be certified and paid pursuant to clauses 60(1)(d) and 60(2)(a).

The fourth function of the Preamble defined by CESMM2 is that it should identify any bodies of open water on the Site. This procedure is explained in the context of paragraph 5.20 later in this chapter.

Paragraphs 5.6 and 5.7 deal with the Daywork Schedule. Three possible procedures are offered. Procedure (*a*) in paragraph 5.6 is use of an *ad hoc* schedule of resources and conditions of payment compiled specifically for a particular contract. Procedure (*b*) is the use of the Federation of Civil Engineering Contractors' daywork schedules with provision for adjustment of rates by percentages inserted by the tenderer. The third procedure, not directly referred to in the CESMM, is not to put a Daywork Schedule in the bill at all. The effect of this is that the FCEC schedules are used without adjustment (clause 52(3)).

Procedure (*b*) should be used unless there are special circumstances which dictate otherwise. It has the advantage that it uses the well-established and well-known arrangement of and conditions attached to the FCEC daywork schedules while allowing the general level of the rates to be subject to the pressures of competition and to reflect differing cost levels in different areas. Materials and plant costs are normally higher in remote areas due to higher transport costs and lower in major conurbations. It is helpful to the financial control of projects if this can be reflected in Daywork rates in the same way as it is reflected in the ordinary rates for work in the bill.

Clause 52(3) of the Conditions of Contract allows for the use of procedure (*b*) only if it is presented in the Bill of Quantities in what may seem an odd way. Unless there actually is a section headed 'Daywork Schedule' in the bill the clause says that the FCEC schedules will be used without provision for any adjustment. To avoid this interpretation, sub-paragraph (*b*) of paragraph 5.6 has to be given in the bill under the heading 'Daywork Schedule'.

Sub-paragraph (*b*) of paragraph 5.6 refers to the adjustment of the labour, materials, plant and supplementary charges proportions of the payment for Daywork. This adjustment consequently applies to the total payment made against expenditure under these four schedules in the FCEC schedules. Thus the percentage addition or deduction stated in the Daywork Schedule in the bill against labour would apply to all payments referred to in schedule 1 of the FCEC schedules. Similarly the adjustments entered for materials and plant would apply to all payments referred to in schedules 2 and 3 respectively. CESMM2 provides for adjustment of Schedule 4—Supplementary charges. The charges referred to in notes and conditions 2(ii), 4 and 7 of schedule 4 are not regarded as supplementary charges for the purposes of CESMM2 because they are actually categories of expenditure covered by the earlier schedules.

During the life of the first edition of the CESMM, there was repeated difficulty due to a small number of contractors who did not understand that the addition or deduction inserted by them in the Daywork Schedule in the Bill of Quantities was in *addition to and not in place of* the percentage additions in the FCEC schedules themselves. In CESMM2, the wording of paragraph 5.6(*b*) has been changed slightly to make this even more clear.

Paragraph 5.6 includes the first use in the CESMM of the word 'inserted'. The CESMM uses the words 'given' and 'inserted' to distinguish by whom words, phrases and other material are put into Bills of Quantities. The word 'given' is always used to refer to material which will appear in the printed Bill of Quantities and which is therefore the responsibility of the compiler of the bill to determine. This includes item descriptions and quantities for Permanent Works. The word 'inserted' is generally used to refer to any material put into the bill by the tenderer and which may differ from one tenderer to another. This includes rates and prices against items, and item descriptions for Method-Related Charges.

Paragraph 5.7 provides for Provisional Sums to be given in the Bill of Quantities—in class A, not in the Daywork Schedule—for the expected expenditure on Daywork labour, materials, plant and supplementary charges. They should be set realistically so that the addition of the percentage adjustments provided under sub-paragraph (*b*) of paragraph 5.6 contributes a realistic amount to the total of the priced Bill of Quantities on which tenders will be compared.

Paragraphs 5.8–5.23 deal with general points about the ordinary bill items which make up the bulk of the Bill of Quantities—Section D: Work items. These items are divided into numbered parts. This is a change from earlier practice when it was common to divide the bill into bills. As the Conditions of Contract and other documents outside the bill itself refer to the bill in the singular, it is consistent to refer to it in the singular within itself. The main division of the bill is into sections which are standardized by paragraph 5.2. Section D (work items) is then divided into numbered parts which differ from one bill to another and are mainly locational or related to the timing of the work. The guidelines for the division of the bill into parts are given in paragraph 5.8 and are sufficiently important to be considered separately.

The criterion of division of the bill into parts is distinction between parts of the work which for any reason are thought likely to give rise to different methods of construction or considerations of cost. Paragraph 5.8 does not say whose thoughts are meant; it must be those of whoever is responsible for the preparation of the bill. Careful consideration of this distinction is very important to the usefulness of the bill to the tenderer during estimating and to all parties in the subsequent financial control of the contract. If well done it makes co-ordination of planning and scheduling with financial control straightforward, allows the prices to reflect properly the realities of the cost of the various parts of work, enables interim valuations to be prepared easily, simplifies agreement of new rates for varied work and encourages prompt settlement of quantities and prices for completed parts of the work in the final account.

If dividing of the bill into parts is done badly or not done at all the estimator may have to set about taking off quantities from the Drawings in order to isolate how much work is where and of what type. Preparation of a programme of work may require further taking off to isolate how much work can be done when and in what sequence. Short cuts in interim measurement cannot be taken by assessing the percentage complete of the various parts of the work. Few bill items are finished with until the end of the job so that preparation of the final account cannot start until then. When a new rate based on an original rate has to be developed and agreed, the cost factors for the work under consideration are difficult to separate. The original rate is a compromise struck from considering several different costs which could only be expressed as one rate. Cash flow forecasting by both the Employer and the Contractor is made difficult by the absence of a breakdown of prices which can be related easily to the construction programme.

Dividing the bill into parts is one of the many aspects of bill preparation and contract financial control generally which require judgement based on knowledge of the factors which influence the Contractor's costs to be exerted by the Engineer and other professional people who may be retained to act on behalf of the Employer. Positive involvement in the acquisition and application of this knowledge is a rewarding aspect of the professional task. It develops positive working relationships with the Contractor's staff, and leads to sounder design decisions, and an enhanced facility to recognize and respond appropriately to efficiency and inefficiency in contractors.

It is intended that the suggestions made in paragraph 5.8 should lead to a greater subdivision of the bill than was normal hitherto. This will enlarge bills to some extent, particularly where similar items recur in several different parts of the bill, but the advantages of subdivision will amply compensate for the enlargement. Estimating will be simplified by the subdivision of quantities into parts which are likely to have different cost characteristics, but will be slightly hampered by the separation of some items which have the same cost into different parts of the job. Planning, valuation of variations, cash flow forecasting and settlement of final accounts are all simplified by thorough subdivision of the bill.

Obviously one of the main purposes of paragraph 5.8 is to enable parts of the bill to be related to operations in a construction programme. This means that planning and estimating can be integrated more easily. This is helpful, but a more easily obtained benefit is its effect on valuation procedures. It has become practice to measure in detail for interim payments only at quarterly intervals, and to use an approximation for the intervening two months. This reduces administrative cost slightly. However, it is more effective to estimate the value of incomplete bill parts on a percentage complete or quantity of main item basis, and to add this each month to the agreed final value of completed bill parts. The site staff responsible for assembling the final account can begin this as soon as work on the first part is complete and need not concern themselves with incomplete parts. The assessment of incomplete parts for interim payment is treated as a by-product of measurements taken for other purposes, such as the Contractor's cost control or bonus scheme.

In theory this produces the complete final account very soon after completion of the work, but in practice there are often reasons why this is not achieved. Some improvement will always derive from the fact that site surveying staff can be instructed to work only on the final account right from the start of their work. Full subdivision of the bill into parts helps this to be realized.

In special circumstances it has been recommended that bills should be subdivided into parts which are synonymous with the activities on a skeletal programme or network for the work specified in the Contract.

Users of the CESMM should not confuse the arrangements for dividing a Bill of Quantities into parts with the division of the CESMM itself into classes. It is not necessary or even advisable to make the items in a bill which are drawn from one class into one part of the bill. Location and timing are the main criteria for subdividing a bill into parts. It is not necessary to have an earthworks bill, a concrete bill, etc. Sometimes location and class are synonymous. The class arrangement is sufficiently apparent from the bill items themselves (whether coded or not) to make use of the class title to separate parts of the bill unnecessary. The exception to this advice is Class A: General items. It is often helpful to make general items a separate bill, in which case Bill Part 1 comprises all the items drawn from CESMM class A and no others.

Paragraph 5.9 gives the rules for interpretation of headings in bills of quantities. It does not say how a heading shall be indicated, but common sense dictates that a heading needs to look different from the ordinary text of item descriptions. Since any text in a heading is read as part of the following item descriptions, any part of a description which is common to a group of items can be used as a heading in order to simplify following item descriptions. This should not be done excessively or where only a small number of items are involved because it inevitably adds extra effort to interpreting individual bill items. The CESMM does not preclude the use of 'ditto' within item descriptions, but it is not encouraged because, like excessive use of headings, it makes the interpretation of individual items difficult and causes problems if items are referred to outside the bill. Difficulties can also arise when new items are added or existing items deleted.

Paragraph 5.10 is one of the most important in the CESMM. It establishes that the bill compiler may elaborate item descriptions and split work into separate items more than the CESMM requires whenever the work to be carried out 'is thought likely to give rise to special methods of construction or considerations of cost'. The paragraph says this 'may' be done, not that it shall be done. Again this is because it would be unreasonable for the Contractor to have a basis for claiming extra payment if the compiler of the bill did not have perfect foresight.

Both paragraph 5.8 and paragraph 5.10 would impose an impossible task on the compiler if they said 'shall' instead of 'may'. As it is, the compiler should recognize the benefit to the administration of the Contract which doing what is suggested in paragraph 5.10 will achieve, and also recognize that he cannot be assailed if he does it less than perfectly. The bill compiler will soon realize that it is in the thorough and comprehensive application of paragraph 5.10 that the exercise of his professional judgement in the preparation of a civil engineering bill of quantities mainly lies.

The CESMM provides a set of rules for describing and itemizing most components of civil engineering work. It is not a strait-jacket procedure; it produces a minimum detail of description and itemization on which the Contractor can rely, but encourages greater detail in non-standard circumstances. Since civil engineering contracts invariably include non-standard work and work in non-standard circumstances, the bill compiler must always expect to amplify the information given so that non-standard characteristics are highlighted. Paragraph 5.10 will normally lead to non-standard itemization and additional description affecting a significant proportion of the items in any bill. The test which should be applied by the bill compiler is to ask himself which unusual features of the work component which he is to describe are significant to the likely cost of the component. If any of these features are not mentioned in the description which would be generated by the application of the CESMM rules they should be mentioned as additional description and the resulting item must be given separately from other items. It is worth bearing in mind that this applies equally to unusually easy and cheap work as to unusually difficult and expensive work.

As this aspect of bill preparation is non-standard, the Contractor cannot rely on its having been done in any particular way. Contractually, by virtue of the wording of paragraph 5.10, he cannot rely on it having been done at all. A bill compiler who does not apply paragraph 5.10 at all may comply with the letter of the CESMM, but he will fail entirely to comply with its spirit and intention. Unusual or differing tolerances are good examples of information which should be given as additional description as a result of applying paragraph 5.10.

When the final account is being prepared, paragraph 5.10 is still in the background. Any extra or varied work should be valued by the Engineer, when applying clause 52 of the Contract, taking full account of any differences in the cost of work due to different locations, different methods of construction and so on. This does not mean, of course, that he should be tempted to introduce price differences in valuing the originally contracted work due to cost differences which were in the original work but were not exposed by the itemization of the original Bill of Quantities. For example, the compiler of the original Bill of Quantities may have decided not to itemize separately a particularly awkward piece of concrete work, although he could have done under the authority of paragraph 5.10 of the CESMM. The Engineer has no right or obligation to review that decision at final account stage if the work has not been varied. The Contractor is not entitled to have an enhanced rate for the part of the work which turned out to cost more than the average for all the work covered by the actual item in the bill. Nor is the Employer entitled to have a reduced rate set for any part of the work which turned out to cost less than average.

Paragraphs 5.11–5.14 deal with some important matters regarding bill item descriptions. Paragraph 5.11 retains the use of the feature of previous standard methods of measurement which produced the main difference between civil engineering and building bills of quantities. It makes it clear that item descriptions only 'identify' work the nature and extent of which are defined by the contract documents as a whole. The item description identifies the work; it does not define the work and it certainly does not aim to contain all the information relevant to pricing the work. If all such information were contained in the item descriptions the Drawings and Specification would be superfluous for tendering.

The main characteristic of civil engineering estimating which distinguishes it from building estimating is that much of the information relevant to pricing is conveyed by drawings and specification. This difference is very deliberately sustained in the CESMM. One reason for this is so that the incentive to complete design before inviting tenders is not weakened; another is that the cost of civil engineering work depends heavily on the shape and position of work and on terrain, and drawings are the best way of indicating shape and position to tenderers.

It is sometimes said that it is unreasonable to expect tenderers to familiarize themselves with a large number of drawings during the preparation of an estimate and that the courts will rule in favour of the Contractor who has missed something shown on a drawing when building up his prices. This is used as an argument for using the full building type of item description in place of the brief identification which is the civil engineering norm. The argument falls down on examination because, if the item description does properly identify the work, it identifies work which, at that stage, is only identifiable on the Drawings and in the Specification. If the bill item does not say enough about the work covered as to make it easy to find the drawing and clauses in the Specification which govern it, then it has not identified the work. The CESMM therefore frequently asks for locational information and/or mark or type numbers in item descriptions, so that identification is made easier.

For these reasons estimators should continue to study the Drawings in order to plan and to cost the operations making up the project. It is helpful if this is made as easy as possible by cross-referencing between the different documents. For example, clauses in the Specification can use the CESMM classes and codes as numbering or referencing systems, and the numbers of Drawings can be given as additional description in bill items or as subheadings in the bill.

The mechanism provided by clause 55(2) of the Conditions of Contract is relevant to the question of the role of the Drawings and of item descriptions. Errors in description in bills and omissions therefrom are corrected and corrections are then treated as variations. For there to be proof of an error there must be a comparison between right and wrong. Anything in a bill which is wrongly treated in relation to the right way shown in the CESMM might be regarded as an error. However, if it is obviously given treatment different from that in the CESMM it could be argued that it is an example of a bill which 'expressly shows' (within the meaning of clause 57) that the CESMM has not been applied so that there was not an 'error' or 'omission' in the bill (within the meaning of clause 55(2)). It is unhelpful to attempt to provide a general solution to this problem. Each case has to be determined on its own facts.

Paragraph 5.12 is a technical statement which allows the descriptions required by the CESMM to be shortened by the use of a reference to the Drawings or the Specification. A reference to a drawing or to a clause in the Specification, if it is to replace description, must be a reference to information which is as precise as the description would be. Suppose, for example, a description of an item for concrete joints read: 'Joint external detail as on drawing 137/11.' This would only identify where the omitted information may be found if drawing 137/11 showed all the particulars of the joint detail required by the CESMM (as per additional description rule A11 of class G) and also did not contain particulars of more than one type of joint detail. If the drawing reference were not specific, the reference to it would have to be made specific by referring to, say, 'detail A' or 'detail C'.

Another requirement placing judgement with the compiler is contained in paragraph 5.13. The CESMM does not set out to define the technical terms which are used in it. The bill compiler is offered no help, for example, in deciding whether the thing he is looking at on a drawing should be regarded as an ordinary concrete structure (measured in classes E, F and G) or a bigger than usual concrete chamber in pipe laying work (measured in classes K and L). No definition is given which establishes the boundary between the application of the two sets of rules. Similarly excavation on a foreshore might be classed as dredging because it is sometimes under water or as general excavation because it is sometimes not under water. A supporting structure for a road sign may be sufficiently substantial to be classed as a metal structure in its own right (measured in class M), or it may be regarded as included in the sign items (measured in class R).

The CESMM deliberately does not make these decisions for the bill compiler. To have attempted to include comprehensive definitions in the CESMM would have led to many arbitrary and misleading classifications with consequent arguments in difficult cases. Instead, paragraph 5.13 requires the compiler to produce item descriptions which eliminate uncertainty by the use of additional description. The paragraph tells the compiler not to worry about into which of the two possible classes an awkward piece of work should go. He should place it in one and then give additional description, peculiar to the item and quantity concerned, which identifies the work precisely in relation to the Drawings or other contract documents. In this way the tenderer can be in no doubt as to which piece of work he is to price against the item, whether the class selected is exactly right or not.

This procedure should be used frequently by bill compilers. There is no virtue in treating classification of work as an intellectual exercise and forcing everything into one or other of the pigeon-holes provided by the Work Classification in the CESMM. To use this approach would make the resulting bill present the work to be priced looking much more normal and standardized than it really is.

If the compiler is unsure which is the right pigeon-hole for a component, he must make sure that the tendering estimators will not also be unsure. Additional description giving location overcomes this problem and ensures that the price is related to the actual work to be done, in whichever class or part of a classification it is placed. It is worth bearing in mind that if the particular work is at one end of the spectrum of work which the ordinary person would consider as of that type, the cost of building it is also likely to be at one end of the cost spectrum and a special price for it would be appropriate in any case.

The CESMM frequently requires items of work which are similar and differ only in one aspect of size to be grouped together into items which cover ranges of that aspect of size. An example is the third division of items in class F dealing with placing concrete in slabs. This groups slabs of different thickness in ranges of thickness as follows

| | | |
|---|---|---|
| 1 | Thickness: | not exceeding 150 mm |
| 2 | | 150–300 mm |
| 3 | | 300–500 mm |
| 4 | | exceeding 500 mm |

All slabs not separately itemized for other reasons whose thickness exceeds 150 mm but does not exceed 300 mm are grouped in a single item which incorporates the descriptive feature 'thickness 150–300 mm'. Paragraph 5.14 says that if all the slabs in this group happen to be of the same thickness, say 250 mm, the item description should state this actual thickness, not the range in which it occurs. This measure ensures that the item descriptions are not needlessly imprecise.

The arrangements for billing work to be done by Nominated Subcontractors in the CESMM do not add much to the procedure set out in the Conditions of Contract in clauses 58 and 59. The items for 'labours in connection' and for 'other charges and profit' are those referred to in clause 59A(5). Sub-paragraph (*a*) of paragraph 5.15 contains a narrowly drawn definition of what is covered by labours in connection with nominated sub-contracts. The definition is slightly amended in CESMM2 by adding the use of temporary roads and hoists and disposal of rubbish. This is intended to align with building practice. It deals differently with nominated sub-contracts which include site work and those which are only for supply of materials or components. The CESMM does not envisage that any bill item should contain both main Contractor's and Nominated Sub-contractor's work. The Prime Cost Items covering nominated sub-contracts are classified as general items in class A. Measurement rule M6 of class A stipulates that any labours in connection with Nominated Sub-contractors other than those defined in the two parts of paragraph 5.15(a) must be identified as 'special labours'. This means that the labours item will be special if it is to include anything that is not in the standard definition or indeed if it is to exclude anything in the standard definition. Additional description rule A6 in class A requires the labours to be included in special labours items to be stated in item descriptions.

Such statements of special labours (sometimes called special attendance) should not be defined vaguely as it is not in the interests of the Employer to require the main Contractor to allow for services the extent of which cannot be foreseen at the main tender stage. Where major special attendance facilities are envisaged but their extent cannot be assessed, they should be made the subject of a Provisional Sum. No provision by the main Contractor of fixing materials for work supplied by a Nominated Sub-contractor can be covered by Prime Cost Items or by the attendance, other charges and profit items.

Paragraph 5.17 establishes that the quantity stated against each item in the original bill should be the quantity which is expected to be required, not a quantity padded to provide a concealed contingency or a small nominal quantity put in to get a price in case some of that work is required. If for any reason some work which cannot yet be defined is to be included in the Contract it must be covered by a Provisional Sum. This rule is based on the sound principle that if the tenderer cannot be told what is required he must not be asked to say what it will cost.

The introduction of clause 56(2) into the Conditions of Contract was a similar development. This clause provides that rates may be increased or decreased if they are 'rendered unreasonable or inapplicable' as a result of the actual quantities being greater or less than those stated in the Bill of Quantities. This applies to all items and means that all prices are treated in the same way as regards their relationship to the stated quantity. Previously items which were indicated as 'provisional quantities' were treated differently from other items. Now that all items are subject to clause 56(2) the term 'provisional quantity' has no significance; it does not appear in the CESMM or in the Conditions of Contract and should no longer be used. Clause 56(2) has the effect that tenderers can safely price each item at a rate which is appropriate to the quantity stated in the bill for that item. It follows that they can safely price the whole of the bill from a comprehensive plan linking all the operations involved. This plan can allow properly for the indirect and time-related costs and be based on a build-up of costs generally which is appropriate to the quantities stated for the total project.

This basis of estimating is essential to sound financial control of civil engineering contracts. Projects now involve a planned group of interrelated operations the cost of which is often dominated by the cost of Temporary Works and specialized plant. Economy depends on achieving planned utilization of Temporary Works, plant and associated labour. The cost of one operation depends on the speed of others; the cost of the whole job depends on the balance of quantities being unchanged. Thus, under clause 56(2) the tenderer is assumed to have planned to do all the work covered by the bill items (excluding Provisional Sums) and to have based his prices on the combination of the quantities actually stated in the Bill of Quantities. Clause 55(1) obliges the Contractor to carry out work in whatever quantities are subsequently required but clause 56(2) means that any change in quantity may give rise to a rate adjustment.

In this situation there is no need for provisional quantities. To designate particular items as provisional quantities could, if the term were so defined, indicate to tenderers that these quantities were more uncertain than the others. However, the items would still be governed by clause 56(2) and the rates set against them would not be treated any differently from the other rates. They could still be adjusted if the quantities changed and tenderers would therefore still have been entitled to price them on the assumption that the stated quantity would be required.

The combined effect of this and paragraph 5.17 is to place a strong incentive on the compiler of the bill to get the quantities as precise as he can. This does not mean measuring accurately from drawings which are tentative; it means measuring accurately from drawings which show the best possible forecast of the nature and extent of the work which will actually be required. It is most important for the Employer and Engineer to work to this philosophy before inviting tenders. The difficulty and uncertainty associated with managing a civil engineering project is strongly related to the incidence of differences between what the Contractor expects to have to do and what he actually ends up doing. ·Clause 56(2) means that the Contractor should be told clearly the nature and extent of the work he must expect to do; any differences between this expectation and later information show up, where previously they could be obscured.

Clause 56(2) was the death knell of the old shopping list approach to civil engineering bills of quantities. Although the Contract Price is arrived at by measuring quantities and valuing them at Contract rates, the Conditions of Contract recognize that construction cost is the total cost of a group of closely interrelated operations. The Employer cannot expect to buy as many or as few of each of the items on the shopping list as he likes at prices per cubic metre or per tonne which apply to any quantity bought.

In the absence of provisional quantities, even the most uncertain quantities have to be estimated as sensibly as possible. At first it may seem strange to put in a bill a quantity for excavation of soft spots or a quantity of hours for pumping plant which is not marked provisional.

The last sentence of paragraph 5.17 refers to paragraph 5.25 which points out that a General Contingency Allowance, if required, should be given as a Provisional Sum in the Grand Summary.

Paragraph 5.18 provides the basic rule for calculating quantities. The simple statement of the general case is that 'quantities shall be computed net from the Drawings'. Only in special cases are quantities measured in a more complicated way than calculating the length, area, volume or mass of the actual extent of the finished work which the Contractor is required to produce. The special cases are those where the CESMM or the Preamble to a particular bill contains conventions for computing quantities. In the CESMM these are in the measurement rules in the various classes of work. They deal with the situations where net measurement is not used, mainly where it would necessitate intricate calculations with little impact on the resulting quantity. Not deducting the volume of concrete occupied by reinforcing steel is a clear example of this type of convention (measurement rule M1 in class F). In older jargon it might have been said that the reinforcing steel was measured 'extra over' the concrete. However, this expression was seldom used in this context. The effect was more clearly and unambiguously achieved by stating the measurement convention. The CESMM avoids the term 'extra over' in order not to raise any ambiguities of this sort. Instead the basic rule of net measurement applies except where another convention is expressly noted. Another example is the measurement of pipe fittings. Measurement rule M3 in class I says that the measured lengths of pipes in trenches include the lengths occupied by fittings and valves. This means that the fittings are measured extra over the runs of pipe in the same way as reinforcement is measured extra over the volume of concrete. The term 'extra over' was often used in this case. The reason that it was seldom used in the previous example is presumably that the accuracy of estimating was such that its use made little difference. Use of the CESMM does not raise such semantic questions. All the items it generates are to be priced to cover the proportion of the cost of the work which is most realistically considered as proportional to the quantity set against each item, bearing in mind any special conventions of quantity calculation and the necessity for covering all costs somewhere.

This approach dispenses with the expression 'extra over'. For example, the piling class provides three items for each group of cast in place piles. Different quantities are set against each: the number of piles in the group, the length of pile material provided and the depth bored or driven. It could be argued that the third item is extra over the second, that the second and third are extra over the first, or even that the first is extra over the other two. Fortunately this argument has no effect on the interpretation of the prices in a Contract.

The last two sentences of paragraph 5.18 are self-explanatory. In the background to both of them is the thought that engineers and surveyors are better employed in almost any other task than in striving for absolute mathematical precision in calculating quantities, whether in original bills, interim accounts or final accounts. Thanks to the central limit theorem of statistics, the accuracy of the sum of a set of numbers is much greater than the accuracy of the individual numbers which make up the set.* The total of the final account is the amount the Contractor gets paid for doing the job and the cost of the job to the Employer. The constituents of the final account total are of no separate importance: they only contribute a lot or a little to that total. The accuracy of the final total is always dominated by the accuracy of the few items which have the largest extended values, either because their quantity or rate is very large. The accuracy of the quantities and rates set against the mass of ordinary and small value items is almost totally insignificant. It was demonstrated by the research work which preceded the preparation of the CESMM that if the 40% of the items in a bill having the least value were taken straight from the tender to the final account without bothering to check whether the quantities had changed or not it would have made a difference to the final account averaging 0·2%. If their value were adjusted in the final account *pro rata* to the change in value of the total of the other larger value items the difference would average 0·03%. This means that if measuring, checking and calculating the actual quantities carried out for that 40% of the items having the least value on a contract of £1 million were to cost more than £350, the cost of doing it would average more than the difference it would make. It still has to be done, otherwise the Employer has an incentive to underestimate the quantities in the original bill for those items which may turn out to be the low value items. It has to be done, but it does not have to be done accurately.

Paragraph 5.19 lists the measurement units which are used against the quantities in bills prepared from the CESMM. The list includes some of the units which are used only within descriptions (mm and $mm^2$). The measurement unit 'sum' is used wherever a quantity is not given against an item. It is equivalent to 'one number' which would be abbreviated to 'nr 1'. Note that the abbreviations for measurement units do not need to have capital initial letters or to be followed by a full stop.

---

* This theorem is better known as the swings and roundabouts effect.

The second edition of the CESMM treats work affected by water very differently from the first edition. The change is seen in paragraph 5.20. Previously the compilers of bills of quantities were required to distinguish work affected by water (other than groundwater) in appropriate item descriptions. This proved to be difficult owing to uncertainty about where to draw the line between work affected by and work unaffected by the presence of a body of water such as a river or canal. Disputes could result if the line was drawn narrowly. If it was drawn widely, the procedure could lose its point. For example, a contract for building a multispan road bridge over a valley involves some work which is 'affected by water'. Any piers founded in the river are definitely so affected, but are the spans which are half over the river and half not? No reasonable person would consider that placing the carriageway over those spans was affected by water. To avoid problems of this sort, it is easy to say in the Preamble to the bill that all the work is affected by water. Unfortunately, this is stating the obvious and achieves nothing.

To avoid this problem CESMM2 simply requires that the presence of bodies of open water (other than groundwater) either on the Site or at a boundary of the Site is to be mentioned in the Preamble to the Bill of Quantities. The Preamble must also give a reference to a drawing which indicates the boundaries and surface level of each body of water or, where the boundaries and surface level fluctuate, the anticipated ranges of fluctuation. It is obviously permissible for the statements about levels and boundaries to be in the Preamble itself where this is more convenient.

Two typical statements exemplifying this rule might be

(*a*) The Site is crossed by the River Corve. The position of the river is shown and the anticipated ranges of fluctuation of its width and surface level are indicated on drawing 137/86.

(*b*) The Site is bounded by the Leominster Canal as shown on drawing 137/87. The width of the canal does not fluctuate but it is anticipated that the surface level may fluctuate between 50·00 a.o.d. and 50·60 a.o.d.

The only type of work for which reference to the presence of bodies of open water is required in the item descriptions themselves is for excavation below water in class E (rules A2 and M7 in class E).

The water surface levels stated in compliance with the last sentence of paragraph 5.20 should be the mean low water level ordinary spring tides and the mean high water level ordinary spring tides of the surface of tidal waters. The presence of ditches and small pools on a site is not intended to call paragraph 5.20 into play. Where they are present, a Preamble clause to this effect may be used.

Paragraph 5.21 of the CESMM uses the definitions of four surfaces from paragraphs 1.10–1.13. The paragraph is self-explanatory once it is read carefully in relation to the definitions. Its intention is to ensure that item descriptions for work involving excavation, boring or driving are clear as regards where the work included in an item starts and finishes. For most such items the descriptions will not mention a Commencing or an Excavated Surface and it will then be assumed that the item covers the full depth from the Original Surface (before any work in the Contract is started) to the Final Surface (when all the work shown on the Drawings has been done). In the special circumstances where an item does not cover the full depth, the intermediate surfaces have to be identified. These will either be the Commencing Surface for the item or the Excavated Surface for the item, or occasionally both. The surfaces do not have to be identified by a level; any identification which is clear and convenient will satisfy the requirement of paragraph 5.21. '250 mm below Original Surface' is an adequate definition of a Commencing or an Excavated Surface, as is '250 mm above formation' if formation is itself clearly defined. In awkward cases it may be necessary to indicate an intermediate surface on the Drawings. The definitions and this paragraph refer to surfaces not levels because they have to cover situations where surfaces are irregular, inclined or vertical, and it would be misleading to apply the word 'level' to such surfaces. Note that CESMM2 includes modified definitions of Commencing and Excavated Surfaces. The effect of these is to make it clear that the upper and lower surfaces of layers of different material occurring within one excavation do not have to be identified as Commencing and Excavated Surfaces.

The last sentence of paragraph 5.21 means that none of the items for excavation or similar work in the CESMM are divided into bands for the volume which occurs between limits of depth below the Commencing Surface. All are given as one item which is classi-

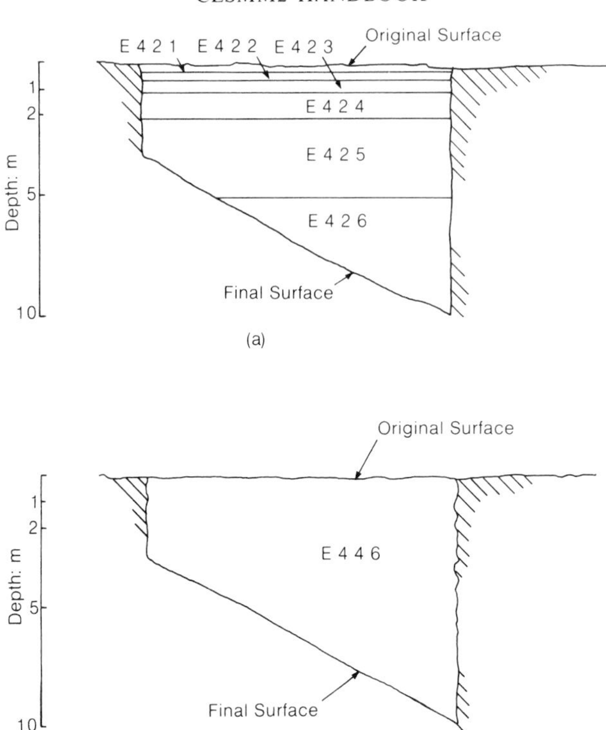


Fig. 7. The CESMM does not divide excavation into depth bands (a), but according to total depth (b)

fied according to a range in which total depth occurs. This is illustrated by Fig. 7.

Paragraph 5.22 suggests that bills should be printed on paper of A4 size with columns ruled and headed in a particular way. This paragraph is not mandatory. The column widths are arranged to suit the requirements of the CESMM when bills are produced using a typewriter or printer with the widest common spacing of ten characters to the inch. A binding margin is accommodated. The maximum occupation of the columns is shown in Fig. 8. The quantity, rate and amount columns each have a capacity of ten million less one. All the work items can be set out on paper of this form. Bills should nor-

| Number | Item description | Unit | Quantity | Rate | Amount £ | p |
|---|---|---|---|---|---|---|
| X999.99 | Description line max 34 characters | sum | 9999999 | 9999999 | 9999999 | 99 |
| | | m2 | 99999.9 | 0.99 | | |

Fig. 8. Column layout described in paragraph 5.22 and typing capacity of the columns when a typewriter producing ten characters to the inch is used

mally contain four headed but otherwise blank pages which are used for the insertion of Method-Related Charges by tenderers. More blank pages may be needed for contracts involving a large number of different operations or divided into many bill parts.

The last four paragraphs of section 5 deal with minor procedural and layout matters to do with the Grand Summary. The Grand Summary does not contain work items and would normally not be on ordinary bill paper. Paragraph 5.25 refers to the General Contingency Allowance which, if required at all, is a Provisional Sum given in the bill and not a percentage to be applied to the total of the work items. Paragráph 5.26 refers to the Adjustment Item which appears as the very last item in the Bill of Quantities. Its function is described in section 6.

There are two general points which deserve mention or emphasis to conclude discussion of section 5.

It is a common practice in the civil engineering profession to prepare the Bill of Quantities for a Contract using the Drawings and the bill for the last similar Contract. This is a perfectly satisfactory procedure provided that the last Contract really was similar in those respects which are cost-significant and provided that the bill which is being used as a model is a good one. Some poor bills have been produced by this method, partly because it discourages adaptation of the bill to the special features of the job in hand, and partly because the model bill may be worse than it looks. Once the handwritten headings, item descriptions and quantities have been well laid out and printed they have an aura of authority, rectitude and precision which may be deceptive—a model bill may have been compiled by a novice from an earlier one compiled by a novice. When the first bill is prepared using the CESMM, this one at least should be compiled from first principles, with perhaps a check back against an older bill only for its coverage of the work. Unless this is done, experience has shown that the item descriptions may be unnecessarily non-standard and the style of the document generally may not be in accordance with the CESMM.

Compilers of bills should keep in mind that a contractor aims to put in as low a tender as he can in order to get work and aims to get as high a final account payment as he can in order to maximize his profit from doing work. A contractor who did not apply any effort or ingenuity to widening the gap between tender and final account would have to be very much more technically efficient than his competitors in order to remain in business. Unfortunately it is often easier to pursue payment than to pursue technical efficiency. Experimentation in the pursuit of payment can only waste hours of professional time. Experimentation in the pursuit of technical efficiency can waste a lot of money in the use of plant, temporary works, materials and labour which turns out to be ineffective.

It is better for the health of the industry and for the service it gives to employers that the pursuit of payment should become less rewarding and the pursuit of efficiency more rewarding. This can be achieved if it is made more difficult to force open a gap between tenders and final accounts. Gap opening feeds only on real and imaginary differences between what the Contractor expected to have to do and what he did have to do; without these there is no gap. The real differences are the variations to the work which the Employer and Engineer know are variations because they recognize that an intention or expectation has been changed. The imaginary differences are those where the Contractor contends that the description in the Bill of Quantities of what he was required to do was misleading or its interpretation uncertain. In order to ensure that his tender is competitive a contractor has to assume that the cheapest interpretation will be acceptable.

The designer has the power to limit the *real* differences by ensuring that his intentions and those of the Employer are as unlikely to change after inviting tenders as is possible in the prevailing circumstances. The compiler of the Bill of Quantities and the Specification has the power to limit the *imaginary* differences by ensuring that the definition of what the Contractor has to do is as clear, unambiguous and extensive as is possible in the circumstances. These two thoughts should be in the forefront of the bill compiler's mind—more prominent even than the general principles and detailed rules of the CESMM itself.

*Schedule of changes in CESMM2*

1. Preamble is to include rules of measurement adopted for any contractor-designed work.
2. Preamble is to refer to clauses in the contract under which the Adjustment Item and Method-Related Charges are to be certified.
3. Preamble is to identify bodies of open water on or at the boundaries of the Site instead of identification in item descriptions for affected work.
4. Daywork schedule alternative (*b*) includes supplementary charges as defined in the FCEC schedules.
5. Definition of labours in connection with Nominated Subcontractors working on Site includes the use of temporary roads and hoists provided by the Contractor for his own use and disposal of rubbish.

# *Section 6. Completion and pricing of the Bill of Quantities by a tenderer*

Section 6 of the CESMM comprises paragraphs dealing with the mechanics of entering rates and totalling a priced bill. The rule for interpretation of the sum, if any, entered against the Adjustment Item is included here so that any sum so entered by a tenderer is understood to be governed by the rules in paragraphs 6.4 and 6.5.

The rules do not explain what the Adjustment Item is for; they only govern how it is applied. Its object is to provide a convenient place at the end of a bill where the tenderer can make an adjustment without altering rates and amounts within the work items themselves. This is often made necessary by the arrival of revised quotations from suppliers or sub-contractors right at the end of the tender period. If a lower quotation for, say, the earthmoving subcontract is received on the day that the tenders are due in, there is not time to adjust all the rates affected, and to recalculate item, page and part totals. The rates and extensions have to be written in the bill and checked, sometimes starting two or three days before the tender is due in. The other reason for last minute adjustment is in relation to the final tender adjudication carried out by the tenderer's senior staff. This process is called different things in different companies; it is essentially a review of the structure and detail of the prices in the tender undertaken in order to assess the risk of not winning the job and the risk of losing money if it is won. These risks apply opposing pressures. Higher prices mean reduced risk of losing money and a greater risk of losing the job; lower prices mean greater risk of losing money and reduced risk of losing the job. Assessment of risk is usually made by discussing the risk issues which the pricing of the work has raised, leading to an agreed level of pricing on which the tender will be based. If this level is not the same as that which the estimator used in his original pricing, the total of the priced bill needs to be adjusted. Again, there may not be time to do this by altering prices: a single adjustment may be needed.

Such adjustments used to be made to the larger prices which were entered against some of the more vaguely worded preliminary or general items in the bill. Using the CESMM no such items are given; each general item is intended to attract a price for a clearly established contractual obligation, whether procedural, managerial or involving site work. It would be quite against the spirit of the CESMM for the rational relationship of price to cost which it fosters to be upset as a result of a legitimate last minute risk adjustment. The CESMM allows, therefore, for an adjustment, made for whatever reason, to be accommodated in the Grand Summary as the last price inserted by the tenderer immediately before the total of the bill. Every rate and price in the bill can be entered and totalled through to the Grand Summary before the final adjudication. The Adjustment Item is mainly a convenience to tenderers; its benefit to Employers is only that it can sometimes eliminate irrational pricing of ordinary general items.

The Adjustment Item could be used as a means of either generally raising or lowering the prices in the bill proper without altering the total of the priced Bill of Quantities. The CESMM does not require the adjustment to be positive. If, to take an exaggerated example, it were negative and equal to 25% of the tender total, the bill rates would be on average 25% above their sensible values. This would make derivation of any new rates difficult and uncertain and generally be unhelpful to the administration of the Contract. It would react strongly to the disadvantage of the Contractor if the quantities in the original bill were over-measured. Equally the insertion of a large positive adjustment would react to the disadvantage of the Contractor if the quantities were under-measured.

Like so many other aspects of the financial control of civil engineering contracts, the Adjustment Item works well if changes to the Contract are modest and if the actual extent of the work required is well predicted in the original bill. It does not pay the tenderer to try to manipulate using the Adjustment Item unless he has foreknowledge of the likely differences between the actual quantities and the billed quantities of work, and he is unlikely to gain such foreknowledge in the short time he has for tendering. If the bill compiler has done what paragraph 5.17 of the CESMM requires him to do, no particular change in quantity will be likely anyway.

The foregoing considerations depend on the Adjustment Item being a fixed lump sum as paragraphs 6.3 and 6.4 of the CESMM dictate. If it were adjustable itself, depending on eventual measured work value, it would not add to the pressure to get quantities right and Contractors would incur little risk by pricing it manipulatively instead of realistically.

Paragraph 6.4 defines how the sum entered against the Adjustment Item is to be dealt with in interim and final certificates. It is worded in terms of clauses 48 and 60 of the Conditions of Contract. Paragraph 6.5 refers to the contract price fluctuations clause* which has been published to give effect to the use of the formula called the Baxter formula. This special condition says that the price fluctuation is calculated by applying indices to the effective value. The effective value includes all contributions to the interim payment which are not based on 'actual cost or current prices'. This definition includes the apportionment of the Adjustment Item in the interim payment. The Adjustment Item sum is subject to fluctuation when the Baxter formula is used as is stated explicitly in paragraph 6.5. In this respect, as in others, the Baxter formula superimposes price fluctuation due to index movement over the adjustments to the Contract sum proper which are assessed using the ordinary provisions of the Contract.

The effect of paragraph 6.4 is that the Adjustment Item is certified in interim payments mainly *pro rata* to the value of the work items (strictly the amount referred to in sub-clause 60(2)(a)) in interim payments until this calculation produces an amount greater than the original Adjustment Item sum in the bill. The original sum is certified at that point and continues to be certified in all subsequent certificates. If the original amount is not previously reached by the *pro rata* process, the original amount is certified in the next certificate issued after issue of the Completion Certificate for the whole of the

* The contract price fluctuations clause (revised January 1979) prepared by the Institution of Civil Engineers, the Association of Consulting Engineers and the Federation of Civil Engineering Contractors in consultation with the Government for use in appropriate cases as a special condition of the Conditions of Contract.

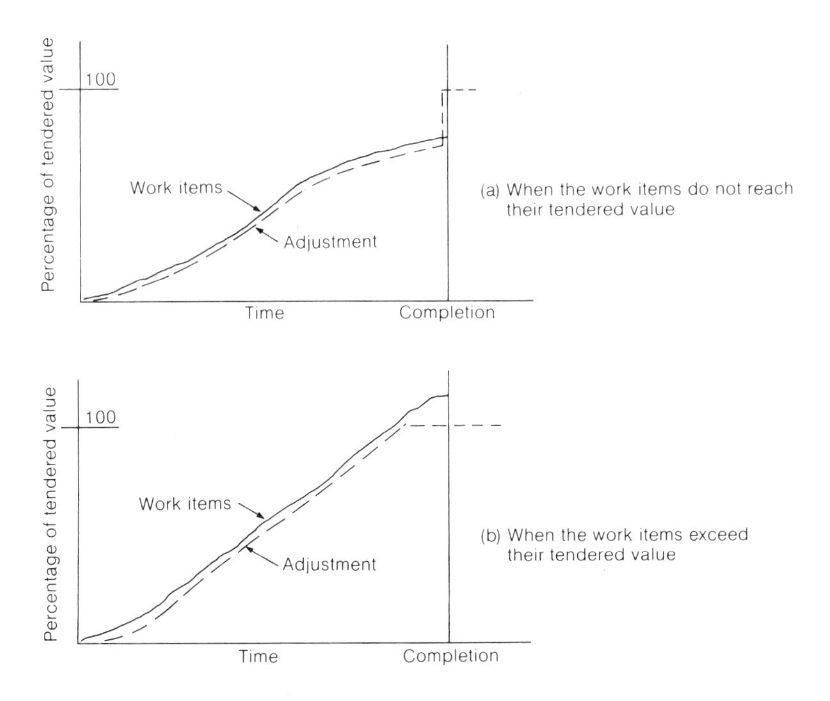


Fig. 9. Time based graph showing how the payment of the Adjustment Item sum relates to the tendered value of work items. It illustrates the effect of paragraph 6.4

Works in accordance with clause 48. In either case, and whatever variations are ordered, the Employer eventually pays and the Contractor receives the original amount shown in the tender against the Adjustment Item. If the adjustment is negative the same procedure applies but to the deduction. The effect of paragraph 6.4 is shown in Fig. 9.

The CESMM itself does not impose any limit on the proportion of the tender sum which may be expressed as the sum against the Adjustment Item. Where the Employer wishes to impose such a limit he can state in his instructions to tenderers that any tender received in which the Adjustment Item exceeds a stated percentage of the tender sum will not be considered.

During the currency of the first edition of the CESMM there were some problems with the settlement of final accounts when contractors had entered substantial sums against the Adjustment Item. On occasions contractors would seek to have the Adjustment Item sum increased when variations increased the value of the work. Alternatively they would seek to have the existence of a substantial positive Adjustment Item taken into account when the Engineer was settling rates for extra work. Neither of these arrangements would be a proper use of the Adjustment Item as paragraph 6.4 makes it clear that the Adjustment Item will appear in the final account at exactly the same value which it had in the accepted tender—whatever had happened to the contract sum in the meantime owing to variations or owing to any other increases or decreases made in accordance with the Contract. Put simply, the Engineer should implement clauses 51, 52 and 56(2) of the Contract as if the Adjustment Item were not there.

This can work very unfairly for the Contractor. For example, on a recent project the Contractor had entered a significant negative Adjustment Item when tendering solely to allow for a late low price arriving from a sub-contractor. He was awarded the contract and shortly afterwards the whole of the particular sub-contractor's work was deleted by a variation order. The Contractor had no contractual basis for asking for the negative adjustment to be deleted as well.

At first sight this seems unfair. Perhaps it is crude rather than unfair because, if the adjustment had been positive, it also should have been left alone, yielding an undeserved bonus to the Contractor in the same way that the actual situation had yielded an undeserved penalty.

This example serves to illustrate an important point about the Adjustment Item which tendering contractors should consider carefully. The mechanism works somewhat crudely and can produce an unexpected bonus or penalty if the work is varied. Contractors who wish to avoid this risk have only one course open to them. They should use strenuous measures to adjust the bill rates themselves before the tender goes in. To use the Adjustment Item for a significant proportion of the tender sum carries the risk of unexpected outcomes in exchange for the right to make a single-figure adjustment at the last moment.

*Schedule of changes in CESMM2*

1. Reference to the Adjustment Item is to be made in the Preamble to the Bill of Quantities.
2. A paragraph has been added to make it explicit that the Adjustment Item is subject to price adjustment as part of the Effective Value when the Baxter formula is in use.

# *Section 7. Method-Related Charges*

### *The purpose of Method-Related Charges*

One of the principal shortcomings of a traditional bill of quantities was that it was only a bill of the quantities of the permanent work left behind when the Contractor's men and machines had moved on. It concealed the contribution to the value of the work made by those men and machines, and by the men who managed them.

This would not have mattered if work was never varied after tenders were accepted, and even then it would not have mattered if the value of the effort of men, machines and managers varied in proportion to the quantities of the permanent work left behind.

Since civil engineering contracts are prone to variation, and since a lot of the Contractor's costs do not vary in proportion to quantity, much of the adjustment of valuation falls outside the simple process of remeasurement. If the Bill of Quantities were only to model price in terms of the quantities of Permanent Works, any reconsideration of costs which did not vary in proportion to quantity would have to be totally unsystematic. The only means of effecting such reconsideration would be by adjustment of rates and by presentation and settlement of claims. Claims bring with them a climate of contention; they invalidate financial control based on forecasts of cost or profit and lead to delay in finalizing payments.

There is no reason why those genuine claims and adjustments arising from necessary changes to the Contractor's methods of working and use of plant and labour should not be within the scope of the semi-systematic processes in the Contract.

To mitigate the claims problem and to draw together the attitudes to construction costs on each side of the contract, the CESMM enables priced bills of quantities to divide the value of work between the quantity proportional elements and the rest.

In commissioning civil engineering work the Employer buys the materials left behind, but only hires from the Contractor the men and machines which manipulate them, and the management skill to manipulate them effectively. It is logical to assess their values in the same terms as the origin of their costs. It is illogical not to do so if the Employer is to retain the right at any time to vary what is left behind and if the financial uncertainties affecting Employer and Contractor are to be minimized.

Tenderers have the option to define a group of bill items and insert charges against them to cover those unexpected costs which are not proportional to the quantities of Permanent Works. To distinguish these items they are called Method-Related Charges. They are themselves divided into charges for recurrent or time-related cost elements, such as maintaining site facilities or operating major plant, and charges for elements which are neither recurrent nor directly related to quantities, such as setting up, bringing plant to site and Temporary Works. If the tenderer enters any such charges, he must use the rules in the CESMM for their classification and basic descriptions.

Since the items required depend on the tenderer's assumptions about method there would be little point in the compiler printing a full range of possible items in the bill. Many would remain unpriced, and descriptions would be too general for effective use of the prices in valuing variations.

Just as there could not be a requirement for a tenderer to price every item in a traditional bill, so there can be no requirement for a tenderer to enter particular Method-Related Charges or to price any if his own interests appear to dictate otherwise.

There was no compunction in a traditional bill to price all items at rates which equalled their anticipated cost plus a uniform profit and overhead margin. The system relied on the tenderer finding it in his own interests to depart from this only to a manageable extent. Similarly Method-Related Charges are only useful in the administration of contracts if the prices entered against them are reasonable in relation to the description of the charge. Reliance is placed on the self-interest of the Contractor to ensure this.

The motives implanted in the Contractor by the procedure must be sufficiently powerful to ensure that, if priced at all, the Method-Related Charges are priced realistically. They should be at least as realistic as prices for measured work—preferably more realistic.

The advantages to the Contractor of pricing Method-Related Charges and of doing so realistically are congruent. They all stem from the fact that a tender so priced matches the subdivision of the total value to be put on the work with the subdivision of the Contractor's anticipated total cost of the work.

If, in compiling his estimate, the tenderer has to consider any part of the cost of assembling and using the labour and plant teams, Temporary Works, supervision and facilities as isolated or time-related sums and not as costs per unit of concrete or excavation, he will consider representing them in his tender as Method-Related Charges. Even if there are no variations to the work, the advantage of a more stable cash lock-up accrues to the Contractor and the Employer when these costs are allowed for in this way.

Interim valuations relate uniformly to expected costs during the construction period. This effect is achieved without the unbalancing of rates which might otherwise be necessary, carrying with it risks of losses if quantities have been wrongly estimated. Similarly the Contractor is protected from losses due to a reduced quantity if the quantity remains within the range to which his original assumptions about methods of work are appropriate. If it goes outside this range, the basis of the assumption and its cost significance are indicated by the Method-Related Charges in the bill and do not have to be re-established.

With variations affecting overheads or increasing the level or duration of provision of a resource or facility, the normal process of valuation includes a review of the Method-Related Charges affected. If there is a variation reducing the work volume, it is seldom necessary for the Contractor to claim for the cost of underutilization of resources or under-recovery of indirect costs. Such costs are covered by Method-Related Charges and payment against them is not reduced. If there is a variation increasing the work volume or causing a delay producing an increase in the costs covered by Method-Related Charges, adjustment of the charges can be made within the terms of the Contract, obviating the necessity for a claim.

This has great advantages: for the Contractor it leads to prompt adjustments, and for all parties it saves administrative effort and time.

Much more than half of agreements on rates for extra and varied work using bills without Method-Related Charges involved consideration of fixed and time-related costs. Many of these claims and new rates would have been unnecessary if prices for the fixed and time-related elements of cost had been realistically separated from the prices for costs which were proportional to quantity.

The simplest justification for Method-Related Charges is provided by the example of the varied quantity. There is a fixed cost in any operation which is the cost of setting it up, of finishing it off and of working up to and down from the steady output of the plant and labour teams involved. This fixed cost is a significant proportion of the cost of an operation unless it is long-term, labour-intensive, only uses small plant and does not require any Temporary Works. The total cost of the operation may be approximated by the function

$$\text{cost} = a + bq$$

where $q$ is the quantity of work in the operation, $a$ is a constant for the fixed cost and $b$ is a constant for the unit cost. The traditional bill of quantities assumes that price is directly proportional to quantity

$$\text{price} = rq$$

where $r$ is the bill rate for the item which covers the operation.

For operations with a low fixed cost this produces a relationship of both cost and price to quantity approximately like that shown in Fig. 10(a). A higher fixed cost produces a narrower range of tolerable quantity variation as shown in Fig. 10(b). If a Method-Related Charge is introduced to allow for the fixed cost, the relationship of price to quantity becomes

$$\text{price} = f + rq$$

The bill rate $r$ is reduced to compensate for the introduction of the fixed Method-Related Charge $f$. This produces the relationship shown in Fig. 10(c). If the fixed charge exceeds the fixed cost by a reasonable margin and the bill rate exceeds the unit cost by the same

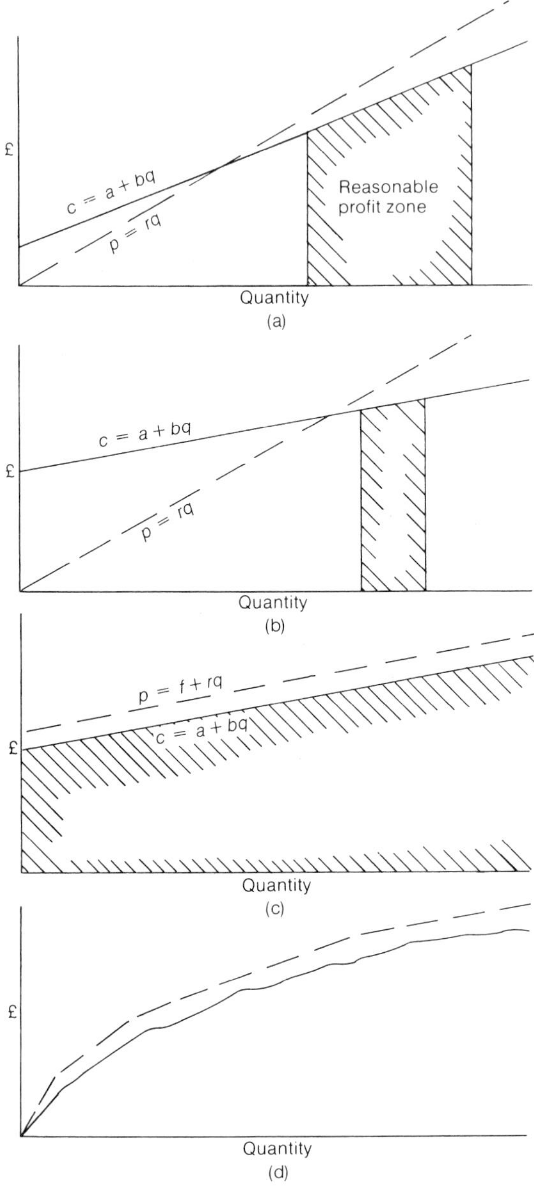


Fig. 10. Price/cost relationships when quantities vary

reasonable margin, the profit on the operation as a whole will exceed the cost by the same margin, whatever the quantity. This is an over-simplification, of course, because the costs of real work do not precisely equal the estimate and do not produce straight lines like those shown in Fig. 10(a)–(c). In reality they produce loosely defined zones of relationship like those shown in Fig. 10(d).

There are three main conclusions to draw from this analysis.

(*a*) Payment for work varied in quantity is more likely to be reasonable and applicable if the prices in the Contract assumed to be proportional to quantity do not have to cover costs which are not proportional to quantity.

(*b*) Claims or the application of clause 56(2) usually ensure that the Contractor does not suffer in the event of changed quantities leading to under-recovery (losses?) on fixed costs. Likewise, separation of Method-Related Charges automatically ensures that the Employer does not suffer from over-recovery on fixed costs when things work out the other way.

(*c*) Method-Related Charges work better than claims or clause 56(2) because the ordinary process of admeasurement and interim certification produces reasonable and applicable payment for work wholly in terms of the original Contract prices. Attention does not have to be diverted from getting the job done to negotiating claims or new rates related to what has already been done.

This third conclusion is idealized as the cost of a civil engineering operation is a function of many factors—some of them totally unpredictable and several of them not simple enough to be represented by recognizable cost parameters.

The principle acknowledged by the procedure for Method-Related Charges in the CESMM does not pursue this approach too far. It recognizes that modern mechanized and technically advanced civil engineering operations involve a significant proportion of cost which is not proportional to the quantity of the resulting Permanent Works. New relationships are embodied in the structure of prices in the Contract to allow for this. Two relationships—those for fixed costs and time-related costs—are added specifically to represent costs to do with methods of and arrangements for carrying out the work. Using a value model with three types of price (a value equation with three types of term) is much more realistic than the traditional bill which recognized only one type of price. The three price bill does not model value perfectly because it does not model all the many variables which affect total cost. It does, however, represent a larger proportion of cost with tolerable accuracy than did the one price bill. It is consequently much more robust as a means of controlling payment in uncertain circumstances and many fewer issues give rise to claims and new rates.

The catalogue of advantages of using Method-Related Charges is so compelling as to suggest that there must also be disadvantages. Comparison of tenders is said to be more difficult, because tenders look different and different policies are adopted for pricing. However, in fact comparison of tenders is easier. Tenders which *are* different also *look* different when Method-Related Charges are used.

When Method-Related Charges are not used, differing policies for pricing and spreading indirect costs are concealed within the wide but tolerated variation between measured work unit rates.

The collateral security for interim payments has been said to be diminished by Method-Related Charges. It is true that Fixed Charges usually mean higher payment at the beginning of the construction period, but later the difference can go either way.

Despite the harmful consequences to the Employer in the very few instances when the Contractor goes bankrupt right at the beginning of the construction period, it would be illogical to deny all the advantages of the procedure to the conduct of all other contracts.

It has been suggested that a carefully worded description for a Method-Related Charge could constitute a qualification to a tender, in the same sense that a qualifying assumption can be stated in a covering letter. This fear is unfounded because statement of an assumption in such a description only records the assumption; acceptance of the tender does not imply that the Employer has underwritten the assumption. By accepting the tender the Employer does not accept the validity of the tenderer's assumptions. For example, if the Employer accepts a tender with reinforced concrete priced at £5 per cubic metre it does not mean that he is liable for the excess if the Contractor finds it cost him more than £5.

For the Employer, the use of Method-Related Charges brings the financial advantage of more predictable final cost; he also benefits from the advantages to his Engineer. The Engineer benefits in a number of ways from method-related costs being brought into the contractual area and out of the area of contention and claim. Also methods can be discussed and administered with the Contractor more openly.

In design the realistic separation of Method-Related Charges from quantity proportional rates makes the rates more reliable as parameters for future projects. Records of Method-Related Charges make it easier to apply knowledge of construction methods to design decisions.

The option of pricing Method-Related Charges will become increasingly popular as users come to recognize its advantages. It is easy to justify the principles of Method-Related Charges, but it is not easy to design practical procedures which give them effect.

Method-Related Charges as introduced in the CESMM were not a totally new concept. The previous standard method of measurement required that the bringing of certain pieces of plant to site should be covered by separate items, which the Contractor might or might not choose to price. The Contractor was expected to put in a separate price for bringing prestressing plant to site—two hydraulic jacks on the back of a Land Rover. It was not really breaking new ground in the CESMM to allow him to separate a price for bringing on a derrick, a batching plant, or a team of tractors and scrapers.

During the life of the first edition of the CESMM, the innovation of Method-Related Charges received patchy utilization. Commonly contractors used the blank pages for Method-Related Charges to enter the sort of 'Contractor's Preliminaries' items which traditional practice had included but did not so as far as declaring items for costs related to construction methods for particular operations. The minority of contractors who did use the new facility more fully were generally pleased with the outcome. There is no doubt that, from the Contractor's point of view, fully declared Method-Related Charges make it easier to secure additional payment related to actual cost when methods and pace of work are changed as a result of variations. It also produces a healthier cash flow. There is no evidence that making a full declaration backfires on the Contractor and those who have tried to use the system fully have usually repeated the exercise.

In the years since Method-Related Charges were introduced, civil engineering work has become more mechanized and faster paced. In times when the demand for work is not buoyant, it may be the Contractor who plans to do the work most quickly who will put in the lowest tender. Making a profit then depends for that Contractor on actually finishing quickly, possibly well inside the period for construction set by the Employer in the Contract. These factors mean that it has become even more important for contractors that the structure of price they are committed to in their priced bills of quantities should realistically represent the fixed costs due to setting up particular construction operations and the time-related costs which they expect to incur. Method-Related Charges give contractors the right and the opportunity to do just this. Not to take up the right and the opportunity is to make a significant addition to the commercial risk which is entirely voluntary and unnecessary.

### *Method-Related Charge procedure*

Bills of quantities compiled in accordance with the CESMM allow space for tenderers to enter both the descriptions and amounts of Method-Related Charges. Descriptions are to be classified as set out in the CESMM but tenderers are at liberty to enter as many or as few charges as they think appropriate.

The charges represent costs which have not been spread over the rates for measured work. Descriptions have to define clearly what the sum entered is to cover so that use of charges in the administration of the Contract can be equitable and uncontentious. The procedure is not obligatory—how much it is used and how realistic is the resulting valuation of work depend solely on the incentive it offers to the users.

Suppose, for example, that a tenderer has chosen to enter Method-Related Charges for the erection and operation of a concrete batching plant. Erection is covered by a non-recurring or Fixed Charge, operation by a Time-Related Charge. Both charges are entered as sums, without quantities, and are not subject to re-measurement. If there are no variations to the work the Contractor will be paid the sums entered, whether or not he uses the batching plant for the length of time that he anticipated. If the work is varied Method-Related Charges are subject to adjustment in the same way as any other rates and prices in the bill, principally as provided by clause 52(2) of the Conditions of Contract. If a variation alters the nature or scope of the Contractor's activity covered by a Method-Related Charge, a revised charge reasonable and applicable in the new circumstances has to be fixed by the Engineer. For instance, the Method-Related Charges for a batching plant might need adjustment because of substantial changes in the concreting operations.

Assessment for interim payment is also prescribed in the Conditions of Contract. The proportion of payment due is that considered by the Engineer to be due (clause 60(2)(a)). The proportion paid should match the proportion of the work covered by each charge which can be substantiated as having been completed by the Contractor. The Engineer will have no grounds for withholding payment against a charge entered for erection of the batching plant if it has been erected on site, and has the capacity or is of the type described.

If the Contractor has entered a substantial but vaguely described charge for bringing on a batching plant, and claims full payment for a cement-caked old 14/10 mixer dragged on to the Site, he should not be surprised to find payment refused.

It is in the Contractor's interest to be explicit in his descriptions of Method-Related Charges if he is to secure timely payment. If events on site can be seen to match a precise description, he will be paid promptly; if events cannot be recognized clearly, then the Engineer may have grounds for withholding payment until the associated Permanent Works are completed.

The mechanism for adjustment to and payment of a Time-Related Charge—such as that for hiring and operating a batching plant—is similar to that for a Fixed Charge. Interim payment should be in proportion to the extent of the activity described which the Engineer is satisfied has been completed.

To assess the proportion of payment due against a Time-Related Charge the Engineer must assess the total period over which the charge should be spread. This period is not required to be stated in the description. It is therefore prudent for the Contractor to ensure that the descriptions and durations applicable to Time-Related Charges match the information given in the programme in the Contract. This provides substantiation for the Engineer's judgement of a reasonable proportion for payment and it safeguards the Contractor against unreasonable apportionment.

In the event of variations, time-related sums become eligible for adjustment if they are rendered unreasonable or inapplicable. Again, the more precisely the work covered by the charges is described, the more realistic can be the adjustment.

Unfortunately realism is not always the goal, and there may be contractors tempted to load or unbalance Method-Related Charges in the hope of circumstances changing in a particular way. There can be no absolute protection against unbalanced Method-Related Charges, as there can be none against unbalanced rates for measured work, but they are a gamble with very long odds. To benefit from unbalanced measured work rates in a traditional bill a contractor has to predict whether particular quantities will increase or decrease. This is difficult enough. To benefit from unbalancing a tender containing Method-Related Charges also requires accurate prediction of the effect of unforeseen variations of the methods and timing of construction. Clairvoyance of a very high order is necessary to achieve any success. Experience shows that loaded Method-Related Charges are more easily recognizable than loaded unit rates. The tenderer's incentive to load is reduced by the more stable cash flow yielded by correctly priced Method-Related Charges.

The procedure for Method-Related Charges has to allow for the situation which develops if the Contractor changes his mind about arrangements which were assumed when pricing his tender. In the event, for example, of a Contractor who has entered charges for a batching plant deciding to use ready-mixed concrete, interim payments against the Method-Related Charges should be certified in proportion to the quantity of concrete placed. It would be unreasonable, in the contractual sense, to do otherwise. If there were variations affecting the volume of concrete placed it would similarly be unreasonable to do anything other than value the varied quantity at a rate carrying an appropriate proportion of the charges for the batching plant.

In the rare situation where there are many variations to the Contractor's plan for the work and also many variations or disturbances to the work itself, the original Method-Related Charges may become almost irrelevant. This can also happen to the ordinary bill rates if the contract goes very sour. No normal form of contract can work well if what has to be done is very different from what both parties expected. As Method-Related Charges represent some of the Contractor's original assumptions, they cannot add to the confusion in these situations; they may help to clarify it.

If the motives implanted by the procedure produced absolute realism, all the Contractor's indirect costs and a substantial proportion of plant, Temporary Works and labour costs would be represented by Method-Related Charges. In theory the measured work rates would cover only material cost and any labour and plant cost which was purely proportional to the volume of permanent materials placed. This split would give a close match between cost and value. A closer match would only be obtained by introducing more categories of charge than the three proposed: the two types of Method-Related Charge and the ordinary measured work rates. In practice, administrative convenience and commercial pressures make it unnecessary to allow separately for method-related costs to quite this extent.

A practical test to use when pricing a bill is to decide which category of charge best represents the way in which a particular element of cost is likely to vary. If cost will vary most closely in proportion to quantity, use measured work rates; if it will vary most closely in proportion to the time for which a set of resources is required, use a Time-Related Charge; if cost is not likely to vary either with quantity or time, then use a Fixed Charge. This means that the costs of the less flexible resources should be partially or wholly covered by Method-Related Charges. Each set of resources, such as major plant items, Temporary Works, site facilities and services, and teams of plant and labour whose composition cannot be varied at will, is likely to involve significant setting up and operating costs which can be represented realistically by Method-Related Charges. There is no precise boundary between the fixed, time-related and quantity-proportional costs of any particular site activity.

The Contractor's interests are best served by making an honest attempt to represent realistically the value of what he has to do, using the three categories of charge available. Representation will never be perfect, but it will be better than that implied by a traditional bill in which construction costs not proportional to quantity were not usually represented.

Since pricing Method-Related Charges is optional to the tenderer, tenders vary from those containing no Method-Related Charges to those with items priced to amount to up to 40% of the tender total and giving a full analysis. Tenderers who choose to use the procedure usually enter 10–20 items amounting to 20–30% of the total tender sum. They cover the usual types of administrative and indirect costs, and extend into more specifically method-related matters such as the provision of haul roads, major plant items, various Temporary Works and the making of formwork. This indicates the range of types of charge which can be expected, all of which are acceptable to the procedure.

The procedure rewards a contractor whose offer contains more realistic and comprehensive information than would traditionally have been disclosed. This can raise the fear in both the Employer and the Contractor that the information so disclosed might be used against them. So it might, but if it is realistic and relevant information, and the use to be made of it is carefully prescribed within the Contract, its exposure can do harm only to a party who would have wrongly benefited from its being concealed. In this respect the procedure could eventually do much to diminish the financial uncertainties surrounding civil engineering construction and to contribute to the equity of its valuation.

Each step in the procedure for using Method-Related Charges requires the Engineer to pay attention to the methods of work both proposed and actually used by the Contractor, and to recognize the origins of their costs. This is not a new requirement and it will only raise difficulties for any Engineers who may have evaded it previously.

That this knowledge is now both required and used is perhaps the most important consequence of the procedure. As construction methods become more mechanized and sophisticated there is a danger of a widening gulf between the designer's and the Contractor's understanding of the cost and value of construction. Use of Method-Related Charges helps to avert this danger so that the separation of design from construction remains competitive with other ways of getting work done.

The procedure for using Method-Related Charges threw up very few problems in the use of the first edition of the CESMM. Despite being a total innovation which requires judgement on the part of Engineer and Contractor, the arrangements have proved robust and relatively free from dispute and uncertainty. For this reason, the rules for using Method-Related Charges are virtually unchanged in CESMM2.

### *Rules for using Method-Related Charges*

The rules for using Method-Related Charges are given in section 7 of the CESMM, supplemented by the classification of Method-Related Charges given in class A. In the event that an Employer wishes to use the CESMM without using Method-Related Charges a statement in the Preamble to the Bill of Quantities to the effect that section 7 of the CESMM does not apply would be adequate. This arrangement should not be sought in normal circumstances as the detailed rules for itemization and description of measured work items are drawn up on the assumption that Method-Related Charges will be used. If they are not used the Employer will be denied the advantages to closer financial control which stem from Method-Related Charges and it is difficult to imagine any circumstances in which the Employer would be well advised not to use them.

Paragraph 7.1 gives definitions of Method-Related Charges and their two subdivisions: Time-Related Charges and Fixed Charges. It should be noted that Method-Related Charges are sums for items inserted by tenderers; they are not rates and the item descriptions for them are not given by the bill compiler.

Paragraph 7.2 amplifies the definitions given in paragraph 7.1. It makes it clear that the tenderer may insert as many Method-Related Charge items as he chooses, that the costs associated with any item will be considered as costs not proportional to the quantities of the other items (mainly the Permanent Works items) and that any costs he allows for within the Method-Related Charges will be assumed not also to be covered by the rates against those items. The converse of these statements is valid: the tenderer need not insert any Method-Related Charges; by this he declares that he anticipates no cost which he wishes to be considered as not proportional to quantities.

This does not mean that the Contractor cannot raise matters which rely on non-quantity proportional costs if valuation of varied work subsequently makes them relevant. However, to do so will invariably seem to be a 'having your cake and eating it' exercise. It is unlikely to be treated sympathetically and with full credulity by the Engineer. It would be like the case of the Contractor who wrote 'included' against a measured item for timber support left in excavations and then claimed extra for having had to leave in much more than he expected. The suspicions are strong that he would not have offered a refund if he had had to leave in less, that he should have allowed for more anyway, and that he probably did allow for more than he subsequently admitted.

The chances of this kind of difficulty being resolved to produce reasonable payment to the Contractor by the Employer are small. Similarly a tenderer may create difficulty for himself by not inserting any Method-Related Charges although he expects to incur significant cost not proportional to quantity. It bears repeating, however, that if the nature and extent of the work required are well predicted in the original bill, these difficulties will be few.

Paragraph 7.2 refers to 'items of work relating to his (the tenderer's) intended method of executing the Works', where 'work' and 'Works' are defined in paragraph 1.5 and clause 1 of the Conditions of Contract respectively. In more colloquial terms it could be said that Method-Related Charges can cover anything which the Contractor will have to do (work) which is a consequence of how he proposes to arrange the construction of the finished product (the Works). 'Method' in this context should therefore be interpreted widely; it includes all activity on or off the Site which the Contractor has to pay for and which is not a direct cost closely related to the quantity of physical components which he will eventually leave behind as the Works.

Paragraph 7.3 urges the tenderer to order his Method-Related Charges and to base his descriptions of the items on the classification for them given in class A, in just the same way as the bill compiler bases the material given by him in the bill on all the other classification tables and notes. Like the bill compiler, the tenderer is not confined to inserting the items listed by the CESMM. He is free to insert other items provided they fall within the definition given by paragraph 7.2. A Method-Related Charge could cover the cost of any peripheral activity—even the cost of the site staff Christmas party (a Fixed Charge?). No practical purpose would be served by this particular example, but it illustrates the point that any indirect cost could in principle be included in a Method-Related Charge.

Paragraph 7.4 is an important rule governing item descriptions for Method-Related Charges. It requires additional description to be given for each Method-Related Charge item. The resulting description is required to be more extensive than the normal bill item description. This is because the item descriptions for Method-Related Charges are not supplemented by the Drawings or Specification; they still identify work, but there is no definition of that work anywhere else in the Contract.

The description against each Method-Related Charge is the statement which determines when it is due to be paid and in what circumstances the amount may be varied. It must therefore describe the extent of the work covered—again 'work' in the sense of activity not in the sense of Permanent Works. Paragraph 7.4 says that the resources expected to be used, if any, and the particular items of Permanent Works or Temporary Works to which the work relates, if any, should be identified. Resources would include teams of plant, gangs of labour and components of temporary works. Thus the general forms of Method-Related Charge descriptions are: 'Fixed Charge for providing/setting up/removing facility/service/resource team comprising . . . to be used for . . .' and 'Time-Related Charge for operating/maintaining facility/service/resource team comprising . . . to be used for . . .'.

The rule does not require the timing or duration of Time-Related Charges to be stated in item descriptions. This is because all Method-Related Charges are sums of money; they are not rates per week or per month. Method-Related Charges cover work the extent of which is a matter at the Contractor's risk. For example, if a tenderer writes in a Time-Related Charge for maintaining a haul road, his item description might be: 'Time-Related Charge for maintaining haul road from A to B during construction of embankment at B'. This item description could be read as equivalent to: 'This is the charge for maintaining the haul road from A to B for whatever time it may take to complete the associated Permanent Works which is the construction of the embankment at B'. It is the words 'during construction of embankment at B' which define the extent of the work covered. This is not a fixed time as the timing and duration of the embanking operation may vary due to circumstances which are incontravertibly at the Contractor's risk. It would be less helpful to say 'for 12 weeks' instead of 'during the construction of the embankment at B' as this would define the extent of the cost allowed for but would not identify the work covered. Contractually it would be as pointless as describing a measured work item as '20 hours of labour, 5 tonnes of structural steel members and 4 hours of a 10 tonne crane' instead of giving a conventional description of the result which it is intended to achieve.

The example bill on pages 115–118 shows typical descriptions of Method-Related Charges. Notice that one activity or piece of work often has both a Fixed Charge for setting up and taking down the resource set or Temporary Works required and a Time-Related Charge for maintaining or operating the resource set. Equally, some work is covered only by either a Time-Related Charge or a Fixed Charge. In practice tenderers often do not separate mobilization from demobilization in Fixed Charges. Where neither end of an activity is indicated, the charge can be assumed to cover both.

Fixed Charges are a way for the Contractor to establish an entitlement to payment for mobilization work when the work is done, i.e. early in the construction period. The CESMM does not prevent the tenderer from allowing for the cost of purchase of resources against Fixed Charges as well as for the cost of bringing them to the Site and setting them to work. A tenderer could allow for the purchase of a piece of plant against a Fixed Charge for mobilization, and give a credit at demobilization for its resale value. This goes further in the direction of helping the Contractor with his cash flow problem than the Employer may wish. It can be prevented, if required, by the inclusion of a statement in the invitation to tenderers that Fixed Charges are not to cover the cost of acquisition of plant or of any other resource which has a significant resale value when it is no longer required for work under the Contract. If a tender is received which appears not to have complied with this instruction it can be rejected. Alternatively, instructions to tenderers may state that tenderers will be given the option of either withdrawing their tenders or adjusting their Method-Related Charges within the original total of the priced Bill of Quantities before acceptance.

Paragraphs 7.5 and 7.8 are emergency provisions—necessary but not expected or intended to be used frequently. Paragraph 7.5 can never affect valuation or conduct of work. It establishes simply that the Contractor is free to arrange and carry out the work covered by the Contract exactly as he chooses within the limits imposed by the Contract. His own descriptions of Method-Related Charges add no constraint; he can change his plans, arrangements and methods without attention to what he may have assumed when tendering.

Paragraph 7.5 is another which has a relevant converse. As the Contractor is free to change his mind, the plans he illustrates by Method-Related Charge descriptions are not necessarily the plans he will adopt and so nothing in a Method-Related Charge description can be construed as a qualification to a tender. If all tenders received are within the terms under which tenders were invited then they are not qualified. All tenderers are given the same opportunity to insert and price Method-Related Charges, so that in that sense also they are not qualifications. It is recommended that invitations to tender should state that no description inserted but not essential by a tenderer against a Method-Related Charge shall be construed as a qualification and that consequently the acceptance of a tender does not constitute an undertaking that the arrangements and methods described or implied are acceptable.

Paragraph 7.6 is crucial to the use of Method-Related Charges. It says that they are not subject to admeasurement. Admeasurement is the word used by clause 56(1) to explain how the quantities of work to be paid for are calculated. Method-Related Charges are not subject to admeasurement; therefore no measurement of quantity and multiplication of rate by quantity is necessary or permitted by the Contract. The sums entered against Method-Related Charges reappear in the final account, unmeasured and therefore not changed merely as a result of the quantity of work carried out being more or less than was originally estimated by the tenderer.

The difference between admeasurement and non-admeasurement can be illustrated. If the quantity of an earth-moving item for Permanent Works is different in the final measurement from that shown in the tender, the amount paid would be adjusted *pro rata* by virtue of the different quantity being multiplied by the original rate. This is the effect of clause 56(1). Conversely, a Method-Related Charge is a sum for an amount of work which is at the Contractor's risk, and whatever amount of work is done the amount paid is the original sum. There is no admeasurement of the amount of time taken to carry out work covered by a Time-Related Charge, or of the amount of resources assembled as the work covered by a Fixed Charge.

Relating paragraph 7.6 with the fact that Method-Related Charges are sums means that the Contractor is entitled to be paid his Method-Related Charges in full for carrying out the extent of the Permanent Works originally tendered for, even if he did fewer or none of the things which were covered by Method-Related Charges. Equally he is not entitled to any more money simply because in the event he did more such work or took longer. To emphasize this point, suppose that a Contractor inserts a Method-Related Charge described as 'Time-Related Charge for dewatering foundation excavation using 64 well points, collecting system, 2 pumps and standby plant'—£65 000. Suppose then that the work is done at the end of a dry spell such that the water table, quite exceptionally, is well below the bottom of the excavation. The Contractor would not even bring any dewatering plant to the Site, much less install it or use it, but he would still be paid the £65 000—not because he would have done any work but because he would have carried the risk of having to do an indeterminate extent of that work. He is entitled to be paid because the Contract places that risk on him and the Employer intends that the Contractor should carry that risk. Having carried it he is entitled to be paid his agreed charge. Only if the risk is varied would the charge also be varied.

Since Method-Related Charges cover work which is at the Contractor's risk, they are priced in the first place according to the nature and extent of that risk. There is no adjustment of Method-Related Charges depending on the outcome of that risk, but there may be an adjustment if the nature and extent of the risk is changed. This is the effect of clauses 51, 52 and 56(2) on the administration of Method-Related Charges. They are not subject to re-measurement but, being part of the rates and prices in the Contract, they are subject to adjustment if they are affected by anything referred to in the Conditions of Contract which brings into effect the procedure of clause 52—mainly variations within the terms of clause 51. Changes in quantity can also lead to adjustment of Method-Related Charges by virtue of clause 56(2). Both clauses 52(2) and 56(2) provide for adjustment of any rate or price in the event of its being 'rendered unreasonable or inapplicable' as a result of a variation or changed quantity. This means that the possibility of adjusting a Method-Related Charge can arise from a variation or change in the quantity or nature of an item in the bill which is not itself a Method-Related Charge. In CESMM2 the fact that Method-Related Charges are prices subject to adjustment under the provisions of clauses 52(1), 52(2) and 56(2) is stated explicitly in paragraph 7.6.

A variation or change in quantity alters what the Contractor is required to do so that in some respect he no longer expects to do what he offered when he undertook the Contract. If the nature and extent of the work required is varied, the original rates and prices may no longer be reasonable and applicable. Clauses 52(2) and 56(2) provide for the adjustment of rates and prices, if necessary, to make them reasonable and applicable in the changed circumstances. Adjustment of prices, method-related or otherwise, comprises correcting them to the same level of reasonableness and applicability as applied to the original prices. Prices should be adjusted by adding or subtracting an amount which is the reasonable extra or reduced cost to the Contractor of doing the work actually required. Reasonable extra or reduced cost is close to actual extra or reduced cost if the Contractor does the work at a level of efficiency close to that on which his estimate was based.

This theoretical consideration is a necessary preliminary to practical illustration of the adjustment of Method-Related Charges. Suppose, for example, a Contract includes a Time-Related Charge for operating a batching plant. A variation to the work may increase or decrease the volume of concrete to be mixed. This will not itself justify any adjustment to the Time-Related Charge which, by virtue of paragraphs 7.1 and 7.2, will specifically not be regarded as proportional to the quantity of concrete required. However, if the variation produces a significant lengthening or shortening of the time for which the operation of the batching plant is required, an adjustment is appropriate: the extent of the work (in the activity sense) is no longer that contemplated in the original price. Thus if the variation adds concrete work which can only be carried out by keeping the batching plant in operation longer than would otherwise be required, an adjustment to the Time-Related Charge based on reasonable cost per extended time period is necessary. Similarly, if the variation deletes the last concrete to be mixed and is issued sufficiently ahead of time to enable the batching plant to be dispensed with earlier; a reduction in the Time-Related Charge is necessary.

This example can be extended to include the possibility of adjustments to Fixed and Time-Related Charges which might cover supervision, site services, Temporary Works and facilities generally. The process of adjustment to Method-Related Charges is never wholly systematic; it is the same as that for adjustment to rates for measured work. It leads to adjustment of the Contract Price which may be very similar to that produced by adjustment of traditional rates which included indirect costs. This adjustment is easier and more systematic. This is due to three factors.

(*a*) Fewer variations lead to rate or price changes. Since rates are not a mixture of costs which depend on different factors, most variations can be valued at Contract prices. The simplest example is that unit rates are applicable over a wider range of quantity variation.

(*b*) More of the adjustment of the Contract price due to variations is based on original prices for work and less on breakdowns of composite unit rates offered after the variation has been ordered.

(*c*) The presence of Method-Related Charges in the Bill of Quantities directs the attention of the Engineer and his staff from the start to observation of the operations which they cover. As these operations are likely to be the subject of price adjustments records are kept of relevant events which take place and the reasons for them.

These three factors together make it much easier, more predictable and more systematic to keep the Contract prices reasonable and applicable. Naturally it is better not to have the variations in the first place, but if they are justifiable, it is better for all parties if they are dealt with by adjustment of Contract prices rather than by using an unsystematic, expensive, unpredictable, tardy and one-sided procedure: claims. A further advantage is that adjustment of prices under clauses 52(1), 52(2) and 56(2) is unequivocally within the duties and responsibilities of the Engineer. There is no justification for the involvement of the Employer's non-engineering staff in the process of adjustment of prices.

Adjustment to Method-Related Charges is permissible only when the procedure for adjustment in clauses 52(2) and 56(2) is applied. There are other clauses in the Conditions of Contract which entitle the Contractor to payment of additional costs of various types. Additional cost must mean cost added to what the cost would otherwise have been. Method-Related Charges may be considered in this assessment only as prices in the Contract related to original assumptions of likely cost.

Time-related Charges are particularly relevant to the assessment of fair or reasonable additional costs due to delays. Assessment is obviously not limited to those matters covered by Time-Related Charges, but they provide a starting point. Extensions of time and additional cost are linked together in several clauses of the Conditions of Contract. Clauses 7(3), 12(3), 13(3), 14(6), 27(6) and 31(2) are examples, although they use various related terms in place of the words 'additional cost'. Application of any of these clauses may involve reassessment of fixed and time-related costs. The practicalities are very similar to those of adjustment of Method-Related Charges under clause 52(2) or clause 56(2), but the technicalities are different. They produce sums claimed under clause 60(1)(d) as additional cost, not as adjusted prices. The effect is much the same except that these clauses only work one way. It should also be noticed that clause 44 itself does not entitle the Contractor to any payment for additional costs which arise for those reasons which entitle him to an extension of time. Any entitlement to costs must be established from another clause.

Where extra work is ordered and no Contract prices apply, the Engineer may accept new Method-Related Charges as part of the procedure specified in clause 52(1).

Daywork payments sometimes have an impact on adjustment of Method-Related Charges. Suppose, for example, that a Contract includes a Time-Related Charge for operating a cableway used for handling materials at a dam site. Suppose also that the Engineer orders some additional work to be executed on a Daywork basis, and that this work involves the use of the cableway. The Contractor is being paid for the total time during which the cableway is provided and operated as a Time-Related Charge, but he is paid for some of it again through the Daywork procedure of clause 52(3) and paragraph 5.6. This may look as though the Contractor is getting paid twice for the same thing, but in reality he is paid only once.

The Contractor makes no allowance in his measured work rates for time spent by plant working on Dayworks and his Method-Related Charges cover only the extent of the work which he judges necessary to complete the ordinary measured work, not the Dayworks as well. If this seems harsh to the Employer, it may console him to recall that, using a traditional bill, the Contractor would have allowed for just the same amount of time-related cost but against the rates. Under the Daywork procedure the Contractor would have been paid just the same amount, but the way in which it would have been presented would not have made it seem as though the Employer was being treated harshly.

Paragraph 7.7 provides that Method-Related Charges shall be certified and paid pursuant to clauses 60(1)(d) and 60(2)(a). This does not add anything to the ordinary criterion of earning payment whereby the Contractor is paid at the prices in the Bill of Quantities for what he has done and is not paid for what he has not done. If all the work covered by a Method-Related Charge has been done the amount of the charge should be certified; if only a proportion has been done, only that proportion of the amount should be certified. CESMM2 requires that paragraph 7.7 should be restated in the Preamble to the Bill of Quantities.

Deciding whether or not payment against a fixed charge is due is usually totally straightforward. By definition a Fixed Charge is made for work which is a one-off activity and it is usually clear at any time whether or not it has been done. If a Fixed Charge is inserted for putting in a haul road from A to B or for bringing an identified plant team to site, the Engineer has no difficulty in ruling that payment is due when he can see that the work covered has been done. One special case is where a Fixed Charge may cover establishment and removal of something. Tenderers should separate these elements into two Fixed Charges, but in the cases where they do not, an apportionment of the charge has to be decided by the Engineer. Another special case is where a Fixed Charge covers a cost which is not associated with something visible on the Site. In such cases the Engineer should ask for evidence that the work has been done before he certifies payment. If, for example, the Contractor has a Fixed Charge for providing insurance additional to that required by the Contract, he should be asked to provide copies of receipts or other papers which show that the premiums have been paid. Engineers should be particularly careful to obtain such evidence before certifying Method-Related Charges for plant assembly or other Temporary Works which, being off the Site, do not vest in the Employer. This might include plant and Temporary Works at a quarry which is opened for the purposes of the Contract but which is located off the Site.

Interim payments for Time-Related Charges are just the same in principle: payment for work done. Since Time-Related Charges usually apply to longer periods of time, they are usually paid in more instalments than Fixed Charges. Again the test of the proportion due is to judge what proportion of the work has been done. In the case of Time-Related Charges this can only mean the proportion or fraction of the total time required for an activity which has already passed. It is not difficult to agree the figure on the top of this fraction—Engineer and Contractor both know when something started—what neither of them knows precisely is how much longer it will go on. After four months of a piling rig being on the site of a harbour extension, the Contractor might claim that this was 4/9 of the eventual time; the Engineer might think it was more realistically 4/12. In any assessment of this type the proportion should not be decided on the basis of the proportion of Permanent Works completed. This cannot be the same as the time apportionment as Time-Related Charges cover costs which are not proportional to quantities.

In practice settling this apportionment is seldom difficult or contentious. The figure on the bottom of the fraction is reviewed every month that the Time-Related Charge is active. It may have to be adjusted from time to time, but it is not worked out from basic principles each time. It is a good technique, and is contractually permissible, for the Engineer to ask for the timing of Time-Related Charges which do not cover the full construction period to be shown on the programme submitted in accordance with clause 14—either as bars on a bar chart or hammocks on a network. This provides a starting point for assessment which can be departed from to the extent that in practice things work out differently from the timing shown on the programme. Contractors who prepare their estimates operationally or with the aid of resourced programmes or networks find that the costs allowed for by Time-Related Charges fall out of the plan for the work anyway without special analysis of the estimate build-up. It is a case of spreading fewer costs back into the rates, not of abstracting more costs from the rates.

Adjusting Time-Related Charges for variations and certifying them for interim payment are reasonably straightforward tasks when considered separately. Doing them together may seem complicated in prospect. In fact it is the clearest realization of the power of the technique to strengthen the financial control of civil engineering projects. It is best to illustrate what happens by an example which uses artificially simplified numbers.

An operation was covered in the Bill of Quantities by a Time-Related Charge of £10 000. The programme submitted in accordance with clause 14 showed that it would begin at the end of month 8 and be finished at the end of month 18—an expected duration of 10 months. For various reasons work proceeded more slowly than the Contractor expected and the work covered by the Time-Related Charge did not begin until the end of month 10. At the end of month 11 the Contractor claimed 1/8 of the Time-Related Charge: £1250. He explained to the Engineer that he expected to recover the two months' delay by the end of month 18, and the forecast duration for the charge was therefore then 8 months. The Engineer considered that the progress during month 11 did not support this forecast and only certified 1/10 of the charge: £1000. During month 12 progress did accelerate and at the end of that month the Engineer was justified in certifying 2/9 or £2222. Certification continued on this basis until, after six months of work at the end of month 16, 6/9 was certified: £6667.

Had there not been a variation at this stage, the later proportions would probably have been 7/9, 8/9 and 9/9 for the next three months. However, during month 17, the seventh month of work, a variation was ordered which added Permanent Works for which the resources covered by the Time-Related Charge would be required. During the month the Contractor advised that this would need the resources to be kept at work for two additional months, making 11 months in all. The Engineer said, 'Wait and see'. At the end of month 17 he certified 7/9: £7778. During month 18 it became clear that the variation would extend the work of the resources affected. The Engineer said that he would vary the charge under the authority of clause 52(2) when he had seen when the work actually finished. He therefore certified 8/9 and 9/9 of the original amount for the charge at the end of each of the next two months. During the tenth month the work continued, at the end of which the Engineer certified no more money against the charge.

The work finished at the end of month 21: eleven months after it had started and two months later than it would have finished without the variation order. The Engineer fixed a new price which was the original £10 000 plus 2/10—yielding £12 000—all of which he certified at the end of month 21. The Contractor thought it might be worth suggesting that the increase should be 2/9, which would have yielded another £222. The Engineer pointed out that this was a wrong interpretation of clause 52(2). The fact that the Contractor would have finished in nine months was his own risk and he had already made an unexpected profit of one month's costs as a result. It was not reasonable to make adjustment so that he made even more unexpected profit on the extra work which was itself unexpected. The Contractor accepted this but made a note to remind the Engineer that this was the correct analysis if a future case arose where the Contractor had taken longer to do something than he expected and made an unexpected loss. The Employer's treasurer, when he saw the 2/10 increase, asked the Engineer, just as a matter of interest, why it had not been 1/10 as the Contractor had had the resources in use only one month longer than the ten months he had expected. The Engineer pointed out to him that the Contractor had had the resources on the Site two months longer than he otherwise would have done. To have claimed back for the Employer the one month's unexpected profit which the Contractor had legitimately made would have been unreasonable because the Contract, to which the Employer had agreed, made it a matter of Contractor's risk how long these resources were required for the original contract work. To have clawed back the unexpected profit on the pretext of a variation would have been as unreasonable as it would have been to compensate his loss in the opposite circumstances.

There are three important points to note from this anecdote. First, the tasks of assessing the Time-Related Charge for interim payments and for adjustment complement rather than complicate each other. The monthly check of what is happening to the likely completion date and why, undertaken in order to decide the proportion to be paid in interim certificates, provides the basis for adjustment of the charge if and when the variation occurs. Second, the final assessment of the adjustment can and does take place immediately after the affected work is complete. Third, the interim payments and the adjustment to the Contract price are wholly in terms of the original prices, of the realities of what actually happened and of why it happened. No claim for prolongation, under utilization of resources or disruption is necessary. This example is idealized to demonstrate the mechanism. In real cases there may be several variations and a multitude of reasons for delay, but however intricate the reality it is easier to resolve when Method-Related Charges are in the Contract than when they are not.

The example assumes that the Contractor worked faster than planned and that the variation added work. There are three other basic cases: the Contractor working slower and a variation adding work, and the two cases of the faster/slower Contractor with the variation deleting work. The outcome in each case depends on whether the variation actually extends or shortens the time for which the resources are in fact required. The approach described in the example produces a sound result in any of the cases. Unlike the claim which might be allowed in the absence of Method-Related Charges, the result may be a reduced cost to the Employer.

The interim and final assessment procedure outlined may seem an added load for the Engineer to carry. In fact it is a lighter load than dealing with the claims which might otherwise result. It is a load which most Engineers are happy to carry as it brings into use the knowledge of construction costs and methods which they may have acquired, and gives their junior staff, helping with and observing the procedure, an opportunity to learn the cost implications of construction methods and timing which they might not otherwise have.

Paragraph 7.5 is a reminder that declaration of Method-Related Charges does not bind the Contractor to follow the plan of work which they represent when he actually comes to do the job. The procedure for interim and final payment against Method-Related Charges must, therefore, allow for the situation where the Contractor is operating in such a way that a particular Method-Related Charge is irrelevant to what is being done. Paragraph 7.8 provides a rule for interim payment in these cases. The Engineer and the Contractor agree how the redundant charges shall be fed into interim payments. The easiest procedure to agree is that the affected charges shall be certified *pro rata* to the proportion of the associated Permanent Works completed. Thus if ready-mixed concrete is being used instead of a site batching plant, any Method-Related Charges for the non-existent batching plant could be paid *pro rata* to the volume of concrete produced. Paragraph 7.8 provides that if no agreement is reached, the redundant Method-Related Charge is added to the Adjustment Item. This has the effect of the sum being paid *pro rata* to the value of all the other items certified but being cut off when the tendered amount is reached (see paragraph 6.4).

The CESMM does not say how a Method-Related Charge is adjusted for variations in cases when the method is changed. Again, the basic provisions of the Conditions of Contract apply. If there are no variations or changes in quantity, all the Method-Related Charges reappear unchanged in the final account; no change to the Contractor's arrangements or methods can lead to a different result.

If there are any variations or changes in quantity the prices in the Bill of Quantities may be rendered unreasonable and inapplicable but the basis of adjustment remains the same; the price should be adjusted to take account of the reasonable change in cost which the variation causes if the Contractor is not to make unreasonable and unexpected profits or losses on varied work. So, in the case of a variation to the concrete quantity after the Contractor has switched to ready-mixed concrete, the redundant Method-Related Charges are adjusted, *pro rata* up or down according to the change in the volume of concrete required. Arithmetically this continues the process which will probably have been adopted for interim certificates, i.e. the Method-Related Charges are spread back over the rates for Permanent Works to which they relate.

If the Contractor is, due to variations, doing work which in detail and extent is quite different from what was expected, and he has completely changed his planned methods and arrangements too, Method-Related Charges are little or no help to the administration of the work. Fortunately things seldom go so wrong. That this is a possibility, however, means that it is wrong to assume that Method-Related Charges are always significantly helpful. They are sometimes little help; in exceptionally badly disturbed contracts they may only help on minor issues. Method-Related Charges are never harmful to control: at worst they become irrelevant; usually they are a positive help.

There are three practical points to note about Method-Related Charges. First, they are subject to the Baxter formula price adjustment because they are a group of prices within the Contract fixed at tender and are consequently included in the effective value as stated in paragraph 6.5.

Second, sub-contractors also have costs which are not quantity proportional. Main contractors should permit any potential sub-contractor to quote using Method-Related Charges to cover his own indirect costs. It will then be up to the main tenderer whether or not he writes the same method-related item descriptions in his tender and carries forward the sub-contract method-related prices, with or without an added profit margin. The popular object of matching main and sub-contract arrangements in order to minimize risk for the main Contractor makes for carrying forward all the sub-contractors' Method-Related Charges. There can be no obligation to do this, however, and some main contractors may suppress Method-Related Charges quoted by sub-contractors by working them back into their rates. Equally, it is permissible for the main Contractor to declare Method-Related Charges for work which will be sub-contracted, even though the sub-contract quotation may not have shown them separately.

Third, the arrangement of Method-Related Charges in the Bill of Quantities is envisaged by the CESMM as a block of items within the general items (class A). Most bills will be split into parts by virtue of paragraph 5.8, and then into subsections of the Work Classification. It is permissible to allow for Method-Related Charges at the beginning of each class within each part of the bill. However, it has been found that it is more convenient to keep them all together in class A bearing in mind that, in accordance with paragraph 7.4, their item descriptions should identify the Permanent Works to which they relate.

*Schedule of changes in CESMM2*

1. Paragraph 7.6 states explicitly that Method-Related Charges are subject to adjustment under clauses 52(1), 52(2) and 56(2).
2. Reference to paragraph 7.7 must be made in the Preamble to the Bill of Quantities.

# *Section 8. The Work Classification*

The Work Classification divides the work which is covered by the CESMM into twenty-five classes lettered A to Y. Each class contains three types of information: an 'includes and excludes' list, a classification table and a table of rules.

The includes and excludes list is given at the head of each class. It tells the user of the CESMM which general types of work are included in a class and which classes cover other similar work which is excluded from that class. In some cases, e.g. in classes A, C, D, I and V, the scope or coverage of the items in a class is also given. These lists should not be regarded merely as an index to the CESMM; they are important to the interpretation of the coverage of the bill items generated by the classes. Clearly the lists do not set out to be comprehensive; they do not mention everything which is included or everything which is excluded.

The classification table is the tabulation of the work components covered by a class, divided into the three divisions as described in section 3. The horizontal lines in the tabulation indicate which lists of features from one division apply to which features in the other divisions. This is usually straightforward, but must be given close attention where the lines are at different levels in the different divisions. Headings are given in some lists of features and are printed in italics in the CESMM. They should be included in item descriptions in all cases where they would not duplicate information. For example, the heading 'Cement to BS 12 or BS 146' is obviously essential in the descriptions for items F 1–3 * 1–4 covering provision of concrete as otherwise the information about cement is not given.

In many places the classification table uses the word 'stated' in phrases such as 'Formwork: stated surface features' and 'Width:

stated exceeding 300 mm'. Written in full these phrases might become: 'This item classification is for formwork which has a particular surface feature. Descriptions of items in this classification shall state the particular surface feature required' and 'This item classification is for things the width of which exceeds 300 mm. Descriptions of items in this classification shall state the actual width of the things required'.

The rules on the right-hand pages are as important as the classification tables. In one sense they are more important as sometimes they overrule the classification table. The rules are arranged alongside the sections of the classification to which they apply. This is indicated by the horizontal lines which align from the left- to the right-hand pages. Rules printed above a double line apply to all items in a class (see paragraph 3.11).

The terms which are printed in italics in the rules are those which are taken directly from the classification table. This style of printing is adopted as an aid to cross-reference between the tables and the rules; it has no effect on interpretation of the rules.

The CESMM uses some untraditional terms. They are adopted to comply with British Standards or to keep up with the move to standardize units and terms under the general umbrella of metrication. Thus pipes have a bore not a diameter, because the bore is the diameter of the hole down the middle, and cannot be confused with the outside diameter of the pipe. Mass is the measure of the quantity of matter; weight is no longer used. The CESMM refers to weight in only one place where it was considered that the alternative phrase 'piece mass' would be totally unfamiliar. The abbreviation for number is 'nr'.

Example bill pages are given in this book for each class in the Work Classification. The examples are not taken from actual contracts. They can be used as a guide to the layout and style of bills and bill items compiled using the CESMM. The example bill items, not being related to a particular job, show less non-standard description amplifying the basic descriptions than is given in real bills. Similarly, in order not to imply that particular specification details are recommended, item descriptions in the example bills frequently refer to hypothetical specification clauses by a clause number or to details on hypothetical drawings. This procedure is permitted in real bills by paragraph 5.12 but it is not adopted in them to the same extent as it has been in the example bill pages.

The examples use the code numbers in the Work Classification as item numbers. This practice is recommended but is not a requirement of the CESMM. It is adopted at the discretion of the bill compiler in accordance with paragraph 4.3.

The examples do not cover all the items which could be generated by the CESMM or even all the items which might occur in one bill. They give hypothetical items which demonstrate those applications of the rules in the CESMM which are novel or would benefit from demonstration for other reasons. The items are laid out as if they were pages from a bill to demonstrate layout, numbering and the use of headings. The examples illustrate the alternative procedures open to bill compilers where the CESMM permits alternatives. Compilers of real bills should try to be consistent, not to demonstrate all the possible alternatives as the examples do.

Each heading and item description in the example bill pages ends with a full stop. This is a helpful discipline when there are two or more headings at the top of one bill page. The full stops help to relate the headings to the lines drawn across the description column which show which items apply to each heading (see paragraph 5.9). Within item descriptions, a semicolon has been used to separate basic from additional description.

# *Class A: General items*

The introduction of the CESMM made the use of general items more purposeful than it was previously. General items include those items which were sometimes called preliminary items in the terminology of the building industry. They are now a group of prices in the Contract which are general in the sense that either they are not related to Permanent Works (such as items for services for the Engineer) or they can conveniently be grouped under a heading which is general (such as items for Method-Related Charges or Provisional Sums).

The CESMM defines cost relationships for general items very precisely. This is an example of the closer relationship of prices to construction costs which the CESMM sets out to achieve. Figure 11 shows how the price/cost relationships for general items are arranged.

In the traditional bill it was assumed that all prices for measured work items should be proportional to quantities, and that prices for general or preliminary items were not necessarily proportional to quantities. The Conditions of Contract make only one reference to preliminary items—in the procedure for valuing work on the outbreak of war. No agreement is assumed in the Contract for any special interpretation of general or preliminary items as regards when they should be paid or in what circumstances varied. This formerly led to hesitance on the part of tenderers when pricing general items, and to uncertainty and contention in the settlement of accounts. A Contractor could have explained that the large sum of money in the general items which was described in very broad terms was mainly for mobilization costs and could have asked for the sum to be paid in the first certificate accordingly. Later that Contract

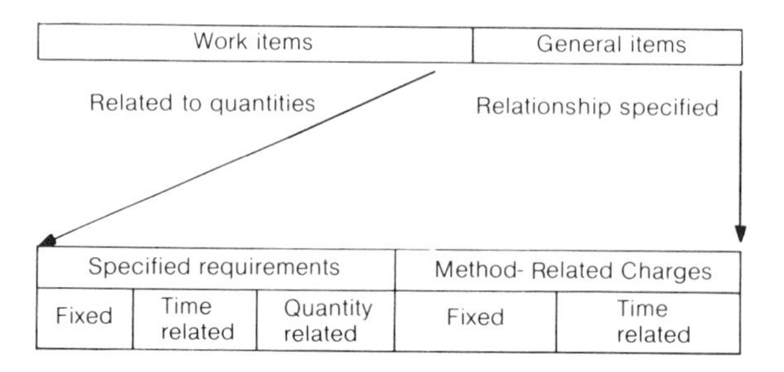


Fig. 11. All prices against items in bills of quantities compiled using the CESMM have an assumed relationship to cost. This relationship is either to quantities which can be observed in the physical work itself (quantity-proportional unit rates), to time (Time-Related Charges) or to neither quantity nor time (Fixed Charges). This figure shows where the items which embody these three relationships are to be found in the CESMM

could have been varied in such a way that the Contractor found it helpful to explain that the sum was mainly to cover the continuing time-related cost of major plant and services. Accordingly it would then have been increased to make it reasonable and applicable to the extended work. Perhaps such a clear cut case has never happend in real life, but that it could happen demonstrates the problem. A traditional bill discouraged tenderers from pricing mobilization costs in appropriate general items because of uncertainty about how the Engineer would include them in certificates. It was safer to allow for such costs against the rates for that measured work which was bound to be done at the beginning of the construction period than to risk that the Engineer would certify only the amount *pro rata* to the value of total measured work.

Figure 11 shows how the CESMM overcomes this problem. All the prices for Permanent Works generated by classes B to Y are either directly related to a measurable quantity of work or are sums related to the extent and nature of a self-contained item. The prices for other work covered by class A are defined as either quantity-related, time-related or fixed. The relationship for each item is stated in its description so that the ordinary processes of interim payment and adjustment specified in the Conditions of Contract can be applied to them rationally and realistically. Prices are controlled more predictably and with a closer relationship to actual cost.

The main division of general items is shown in the first division descriptive features of class A. The bill compiler should give items for all the obligations required by the Contract and all the services which the Contractor will be required to provide, presenting their descriptions in conformity with the notes and the general requirements in sections 1–5 of the CESMM.

The insurance items A 1 2–4 0 cover only insurance which is a contractual requirement: the two risks noted in (a) and (b) of clause 21 and the insurance to be effected under clause 23. They do not include a special item for insurance against the necessity for repair or reconstruction of any work constructed with materials and workmanship not in accordance with the requirements of the Contract. Such items are not normally provided and no provision for them is made in the CESMM. The Contract may require the Contractor to indemnify the Employer against the consequences of other risks, but as he is not necessarily required to take out insurance to cover such indemnities, no items are given for insuring them. As the note at the foot of page 17 of CESMM2 points out, a tenderer may insert Method-Related Charges for any such insurance if he so wishes.

Rule D1 is an important rule which determines when an item should be given in the bill for a specified requirement. Its effect is broadly that if the Specification or any other contract document states a requirement that the Contractor should do something in a particular way or to a particular extent then it should be given as an item for separate pricing. This rule has a dual function. It ensures that the tenderer's attention is drawn to anything about which he might normally expect to make his own decisions but which in this case is pre-determined in the Contract. Temporary Works and plant illustrate this point. If the Contractor is required to construct a cofferdam or to provide compressed air plant for tunnelling, items for them must be given, and relevant details stated if they are prescribed. If the Contractor may decide for himself whether to build a cofferdam or to use compressed air, no items are given as specified requirements. A tenderer is free to insert Method-Related Charges for this work if it is not a specified requirement. This is why Temporary Works appear in the classification table against both specified requirements (A 2 7 *) and Method-Related Charges (A 3 5–6 *).

The other function of specified requirements items is to provide a price which can be adjusted in the event of variation. If a Contract does not expressly require something to be done in a particular way it is unlikely that it will be changed. If something is a requirement, particularly if it is a detailed requirement, such as the provision of a cofferdam which has been designed by the Engineer, it may be changed. If it is changed it is helpful to have a price in the Contract for it, so that its cost is not spread diversely among other prices. The bill compiler should go through the Specification looking for specified work of this type, particularly when compiling a bill using the CESMM for the first time.

Rules M2 and A2 are the rules by which the cost relationships of specified requirements items are shown in the bill. Items covering fixed costs will normally be sums not subject to quantified measurement. They will usually be described by phrases like 'Establishment and removal of . . .'. Time-related cost items will be either quantified, measured in hours or weeks, or given as sums. The CESMM does not say when quantities or sums shall be used; this judgement is left to the compiler. The example bill for class A shows some items for time-related costs quantified (e.g. A 2 2 1.1, A 2 4 2 and A 2 4 3) and some not (e.g. A 2 1 1.2, A 2 1 2.2 and A 2 3 3.2). An appropriate procedure to adopt is to quantify only when the cost is directly proportional to a quantity which can be measured and which is under the control of the Engineer. This would mean measuring in weeks such work as providing vehicles and staff, but leaving as sums less regular work such as maintaining offices and laboratory equipment for the duration of construction.

Items for plant which is classed as specified requirements should be given in two items to comply with rule A2. These are items for operating or maintaining and establishment and removal. Rule A2 does not require all specified requirements to be divided between two items. Quantified items are not usually for 'services or facilities' and do not need to be split in this way. Plant items needing two items are exemplified by the items for pumping and dewatering shown in the example bill (A 2 7 6.1–2 and A 2 7 7.1–2).

They would have been extra over only if there had been a note to the effect that operating was measured while plant was standing by. Rule A2 refers to services and facilities. It is therefore not necessary to give establishment and removal items for work which, although a specified requirement, is not a service or facility and has negligible establishment and removal cost. Testing items need to be considered carefully in this context. If tests of a particular type of work do not involve establishment of testing facilities it is not necessary to give an item for it. The items in the example bill for pipe testing (A 2 6 0.1–4) exemplify this. Tests involving assembly or construction of cost-significant testing facilities would be treated differently. Testing items generally should be given full descriptions or equivalent references to the Specification. Samples involving site work should be regarded as tests, but normal samples of materials are seldom sufficiently costly to warrant separate billing.

CESMM2 does not require separate items to be given for specified plant standing by. This is a manifestation of a pervasive change in CESMM2 which is that almost all the items in the various classes for payment for delays, standing by and similar activity (or inactivity) measured by the hour have been eliminated. This is because of problems with the establishment of what is deemed to be covered by such items. Is it the consequential costs of a long delay of some days or weeks or of a short delay lasting less than half a shift? Clearly the cost per hour is very different at these extremes. To avoid these problems, CESMM2 does not require items to be given for standing by, delays, etc. If any of these occur for reasons which the Contract says entitle the Contractor to extra payment, the mechanisms in the Contract must be used for assessing the payment, not predetermined rates in the Bill of Quantities.

The classification table in class A lists the elements of possible Method-Related Charges items which can be expected to occur frequently. Contractors should note that the rules for Method-Related Charges in section 7 of the CESMM place no constraint on the type of expense which can be covered by a charge if the Contractor considers that this would be a sensible expense to show separately when he is compiling his tender. Contractors should therefore enter items for Method-Related Charges not shown in the classification to cover the less common expenses which it is considered will not be incurred in proportion to the quantities of the Permanent Works. These might include, for example, additional insurances, financing charges and royalties to owners of off-site borrow pits. It would be permissible to show head office overheads and profit separately as Method-Related Charges but it is unlikely that tenderers will consider it in their interests to do so.

Employers and their Engineers may find it strange that the Contractor's own interest should be the only criterion for pricing Method-Related Charges. However, even in traditional bills any breakdown of the total contract sum between the Permanent Works items was only adopted by the tenderer in his own interest and it would be false to assume that different considerations apply to Method-Related Charges.

A tenderer cannot be forced to price each item in a bill to cover all the anticipated costs associated with that item and no others, and to contain a general risk and profit margin uniform with all the other items. The prices are most helpful for control of the Contract if he does price in this way, but it cannot be insisted on or assumed. The CESMM relies on commercial motives working through the Conditions of Contract to make rational pricing in the tenderer's interest as well as the Employer's. Method-Related Charges, the Adjustment Item and the details of Permanent Works itemization are all intended to encourage this. The effect of all of them can be nullified if the tenderer discovers that the quantities given in the bill are not those most likely to be required.

Tenderers enter Method-Related Charges in the space provided in the general items part of the bill. The example bill contains a greater variety of Method-Related Charges than would normally be found in any one tender. They appear in manuscript to emphasize the fact they are not given in the bill by the compiler. The items show a wide range of the possible insertions. No tender is likely to have the particular combination of Method-Related Charges shown. Note that the descriptions use a variety of ways of identifying the resources used and the Permanent Works, if any, to which they relate. The examples demonstrate that it is not difficult to be unambiguous using short descriptions. Note also how the extent of the work covered by Time-Related Charges is indicated by reference to 'the duration of construction', particular events such as 'completion of the frame' or particular sections of the work such as 'concreting operations at main treatment works site'. These are the most useful definitions of the anticipated timing: more helpful to the Contractor and Engineer than a statement of duration unrelated to other things. 'Duration of construction' is used in the examples in preference to 'contract period' as the latter term implies that the cost will be incurred for a fixed duration whether construction takes a shorter or longer time than the period for completion stated in the Contract.

The Provisional Sums begin with the Daywork items, classified on the assumption that the alternative form of Daywork Schedule specified in paragraph 5.6(*b*) is included as section C of the Bill of Quantities.

Although the classification mentions telephones for the Engineer's staff at A 2 2 2, calls made are usually allowed for as a Provisional Sum at A 4 2 0.

Prime Cost Items are grouped at A 5 * 0 and A 6 * 0. Rule M6 requires that additional description shall be given if the labours to be provided are not precisely those listed in paragraph 5.15 (*a*). This description must state that the labours are special labours and give a full definition of the work required, including all details of the work and the stages in construction at which it will be required. The tenderer has to give a firm price for the work and therefore needs to be able to assess precisely what plant, labour and Temporary Works will be required, for how long and when. 'When' is important because it will affect which resources are then available and in position for other purposes and which have to be provided specifically for the special labours.

*Schedule of changes in CESMM2*

1. Items for standing by plant are not given.
2. Supplementary charges are added to daywork items.
3. Prime cost items are rearranged.
4. Separate items for specified requirements carried out after completion.

| Number | Item description | Unit | Quantity | Rate | Amount £ | p |
|---|---|---|---|---|---|---|
| | PART 1: GENERAL ITEMS. | | | | | |
| | Contractual requirements. | | | | | |
| A110 | Performance bond. | sum | | | | |
| A120 | Insurance of the works. | sum | | | | |
| A130 | Insurance of constructional plant. | sum | | | | |
| A140 | Insurance against damage to persons and property. | sum | | | | |
| | Specified requirements. | | | | | |
| A211.1 | Establishment and removal of offices for the Engineer's staff. | sum | | | | |
| A211.2 | Maintenance of offices for the Engineer's staff. | sum | | | | |
| A211.3 | Maintenance of offices for the Engineer's staff after issue of the Completion Certificate. | wk | 20 | | | |
| A212.1 | Establishment and removal of laboratories for the Engineer's staff. | sum | | | | |
| A212.2 | Maintenace of laboratories for the Engineer's staff. | sum | | | | |
| A221.1 | Transport vehicle; as Specification clause 184.8. | wk | 208 | | | |
| A221.2 | Transport vehicle; as Specification clause 184.8 for use after issue of Completion Certificate. | wk | 20 | | | |
| A229 | Set of progress photographs comprising six prints. | nr | 200 | | | |
| | | | PAGE TOTAL | | | |

GENERAL ITEMS

| Number | Item description | Unit | Quantity | Rate | Amount £ | p |
|---|---|---|---|---|---|---|
| | Specified requirements. | | | | | |
| | Equipment for use by the Engineer's staff. | | | | | |
| A231.1 | Establishment and removal of office equipment. | sum | | | | |
| A231.2 | Maintenance of office equipment. | sum | | | | |
| A231.3 | Maintenance of office equipment after issue of Completion Certificate. | wk | 20 | | | |
| A232.1 | Establishment and removal of laboratory equipment. | sum | | | | |
| A232.2 | Maintenance of laboratory equipment. | sum | | | | |
| A233.1 | Establishment and removal of surveying equipment. | sum | | | | |
| A233.2 | Maintenance of surveying equipment. | sum | | | | |
| A242 | Attendance upon the Engineer's staff, chainmen. | wk | 104 | | | |
| A243 | Attendance upon the Engineer's staff, laboratory assistants. | wk | 104 | | | |
| A250 | Testing of materials; concrete test cubes; samples and methods of testing as Specification clauses 186.1 to 186.9 | nr | 400 | | | |
| | Testing of the Works. | | | | | |
| A260.1 | Clay pipes nominal bore not exceeding 200 mm, test as Specification clause 187.1. | sum | | | | |
| A260.2 | Clay pipes nominal bore 200 - 300 mm, test as Specification clause 187.2. | sum | | | | |
| | | | | PAGE TOTAL | | |

GENERAL ITEMS

| Number | Item description | Unit | Quantity | Rate | Amount £ | p |
|---|---|---|---|---|---|---|
| | Specified requirements. | | | | | |
| | Testing of the Works. | | | | | |
| A260.3 | Spun iron pipes nominal bore not exceeding 200 mm, test as Specification clause 187.3. | sum | | | | |
| A260.4 | Spun iron pipe nominal bore 200 - 300 mm, test as Specification clause 187.3. | sum | | | | |
| A260.5 | Aeration tanks; watertightness test as Specification clause 188.1. | sum | | | | |
| A260.6 | Final settling tanks; watertightness test as Specification clause 189.1. | sum | | | | |
| A260.7 | Storm tanks; watertightness test as Specification clause 190.1. | sum | | | | |
| | Temporary Works. | | | | | |
| A271.1 | Traffic diversions. | sum | | | | |
| A272.1 | Traffic regulation; establishment and removal. | sum | | | | |
| A272.2 | Traffic regulation; continuing operation and maintenance. | wk | 104 | | | |
| A273.1 | Establishment and removal of access roads. | sum | | | | |
| A273.2 | Maintenance of access roads. | wk | 104 | | | |
| A276.1 | Establishment and removal of pumping plant. | sum | | | | |
| A276.2 | Operation and maintenance of pumping plant. | h | 500 | | | |
| A277.1 | Establishment and removal of de-watering plant. | sum | | | | |
| A277.2 | Operation and maintenance of de-watering plant. | wk | 17 | | | |
| | | | PAGE TOTAL | | | |

GENERAL ITEMS

| Number | Item description | Unit | Quantity | Rate | Amount £ | p |
|---|---|---|---|---|---|---|
| | Method-Related Charges. | | | | | |
| | Accommodation and buildings. | | | | | |
| A311.1 | Set up offices ; Fixed. | SUM | | | | |
| A311.2 | Maintain offices for the duration of construction ; Time-Related. | SUM | | | | |
| A311.3 | Remove offices ; Fixed. | SUM | | | | |
| A314.1 | Set up stores and materials compound ; Fixed. | SUM | | | | |
| A314.2 | Remove stores and materials compound ; Fixed. | SUM | | | | |
| A315.1 | Set up canteens and messrooms; Fixed. | SUM | | | | |
| A315.2 | Operate canteens and messrooms for the duration of construction; Time-Related. | SUM | | | | |
| | Services. | | | | | |
| A321 | Set up electricity supply and standby generator ; Fixed. | SUM | | | | |
| A322 | Water supply for the duration of construction ; Time-Related. | SUM | | | | |
| A325 | Site transport for the duration of construction comprising one 5t truck and two tractors and trailers ; Time-Related. | SUM | | | | |
| A327 | Welfare facilities complying with HSAW regulations for the duration of construction ; Time-Related. | SUM | | | | |
| | 1/5 | | Page total | | | |

GENERAL ITEMS

| Number | Item description | Unit | Quantity | Rate | Amount £ | p |
|---|---|---|---|---|---|---|
| | Method-Related Charges. | | | | | |
| | Plant. | | | | | |
| | 35t crane for excavation and concreting of main tanks. | | | | | |
| A331.1 | Bring to Site ; Fixed. | sum | | | | |
| A331.2 | Operate ; Time-Related. | sum | | | | |
| A331.3 | Remove ; Fixed. | sum | | | | |
| | Team comprising two pusher/ripper bulldozers, six motor scrapers and one spreader bulldozer for main embankment shale filling. | | | | | |
| A333.1 | Bring to Site twice and remove twice; Fixed. | sum | | | | |
| A333.2 | Operate ; Time-Related. | sum | | | | |
| A339.1 | Bring to Site and remove jacks and other prestressing plant for stressing bridge 12 deck beams ; Fixed. | sum | | | | |
| A339.2 | Bring to Site and remove 3 grout pans and 2 grout pumps for work in items 2.B111.1 to 2.B345 inclusive ; Fixed. | sum | | | | |
| | 1/6 | | Page total | | | |

GENERAL ITEMS

| Number | Item description | Unit | Quantity | Rate | Amount £ | p |
|---|---|---|---|---|---|---|
| | Method-Related Charges. | | | | | |
| | Temporary Works. | | | | | |
| | Road diversion at Newton Street during construction of culvert 21b. | | | | | |
| A351.1 | Install ; Fixed. | sum | | | | |
| A351.2 | Operate and maintain ; Time-Related. | sum | | | | |
| A351.3 | Remove ; Fixed. | sum | | | | |
| A353 | Install access road, site entrance to batching plant near downstream tunnel portal ; Fixed. | sum | | | | |
| A356 | Pumping as required during excavation and concreting of tower foundations ; Time-Related | sum | | | | |
| A361.1 | Erect scaffolding surrounding administration block ; Fixed. | sum | | | | |
| A361.2 | Hire of scaffolding surrounding administration block from completion of frame until completion of brickwork ; Time-Related. | sum | | | | |
| | Erection of falsework to support bridge deck formwork ; Fixed. | | | | | |
| A362.1 | Bridge 1. | sum | | | | |
| A362.2 | Bridge 2. | sum | | | | |
| A362.3 | Bridge 3. | sum | | | | |
| A362.4 | Bridge 4. | sum | | | | |
| | 1/7 | | | Page total | | |

GENERAL ITEMS

| Number | Item description | Unit | Quantity | Rate | Amount £ | p |
|---|---|---|---|---|---|---|
| | Method-Related Charges. | | | | | |
| | Temporary Works. | | | | | |
| A363 | Temporary sheet pile wall to retain excavation on South side of Long Lane diversion from chainage 125 to 275; Fixed. | sum | | | | |
| A364.1 | Make 4 sets of full height forms for treated water reservoir columns; Fixed. | sum | | | | |
| A364.2 | Make 120 raked side panels and 20 rectangular end panels for formwork to 3m high lifts in main dam concreting; Fixed. | sum | | | | |
| A364.3 | Use 15 m long circular steel form for lining diversion tunnel; Time-Related. | sum | | | | |
| | Supervision and labour. | | | | | |
| A371.1 | Management and supervision for the duration of construction; Time-Related. | sum | | | | |
| A371.2 | Additional management and supervision during construction of the main pumphouse; Time-Related. | sum | | | | |
| A372 | Administration for the duration of construction; Time-Related. | sum | | | | |
| A373.1 | Labour for maintenance of plant and site services during earth moving and concreting operations; Time-Related. | sum | | | | |
| A373.2 | Labour for concreting gang during concreting operations at main treatment works site; Time-Related. | sum | | | | |
| | 1/8 | | Page total | | | |

GENERAL ITEMS

| Number | Item description | Unit | Quantity | Rate | Amount £ | p |
|---|---|---|---|---|---|---|
| | Provisional Sums. | | | | | |
| | Daywork. | | | | | |
| A411 | Labour. | sum | | | 50000 | 00 |
| A412 | Percentage adjustment to Provisional Sum for Daywork labour. | % | | | | |
| A413 | Materials. | sum | | | 25000 | 00 |
| A414 | Percentage adjustment to Provisional Sum for Daywork materials. | % | | | | |
| A415 | Plant. | sum | | | 25000 | 00 |
| A416 | Percentage adjustment to Provisional Sum for Daywork plant. | % | | | | |
| A417 | Supplementary charges. | sum | | | 20000 | 00 |
| A418 | Percentage adjustment to Provisional Sum for Daywork supplementary charges. | % | | | | |
| | Other Provisional Sums. | | | | | |
| A420.1 | Permanent diversion of existing services. | sum | | | 20000 | 00 |
| A420.2 | Landscaping. | sum | | | 25000 | 00 |
| A420.3 | Repairs to existing tanks. | sum | | | 15000 | 00 |
| A420.4 | Repairs to existing pipelines. | sum | | | 10000 | 00 |
| | | | | PAGE TOTAL | | |

GENERAL ITEMS

| Number | Item description | Unit | Quantity | Rate | Amount £ | p |
|---|---|---|---|---|---|---|
| | Nominated Sub-contracts which include work on the Site. | | | | | |
| A510.1 | Electrical installation. | sum | | | 50000 | 00 |
| A520.1 | Labours. | sum | | | | |
| A540.1 | Other charges and profit. | % | | | | |
| A510.2 | Flow recording and control equipment. | sum | | | 25000 | 00 |
| A520.2 | Labours. | sum | | | | |
| A540.2 | Other charges and profit. | % | | | | |
| A510.3 | Pumping machinery. | sum | | | 42000 | 00 |
| A530 | Special labours; attendance on commissioning and testing as Specification clause 207.1, carried out upon completion of pump installation. | sum | | | | |
| A540.3 | Other charges and profit. | % | | | | |
| | Nominated Sub-contracts which do not include work on the Site. | | | | | |
| A610.1 | Precast concrete bridge beams. | sum | | | 37000 | 00 |
| A620.1 | Labours. | sum | | | | |
| A640.1 | Other charges and profit. | % | | | | |
| A610.2 | Precast concrete filter wall units. | sum | | | 12000 | 00 |
| A620.2 | Labours. | sum | | | | |
| A640.2 | Other charges and profit. | % | | | | |
| | | | | PAGE TOTAL | | |

# *Class B: Ground investigation*

The CESMM includes rules of measurement for ground investigation which can be used either when the work is part of a larger construction contract or when it is a self-contained ground investigation contract. The rules are not complicated. Item descriptions generally identify a sample or test and are measured by the number of samples or tests required. Descriptions are linked to relevant British Standards. In the second edition of the CESMM, the measurement of ground investigation has been overhauled thoroughly.

The items for trial holes and boreholes have been changed significantly and are governed by the classification table B 1–3 * * and the accompanying rules. There are at least two items for each type of hole. There is one item for the number of holes and various items for their depth. The numbered item is intended to cover costs proportional to the number of holes (such as those for moving boring rigs from one position to another and setting up to excavate a hole). The depth items for trial holes distinguish between depths excavated in rock and other materials, depth supported and depth backfilled. The number items are subdivided according to whether rock excavation is included or not and according to the maximum depth of hole. The itemization of trial holes takes into account the fact that the nature of the work cannot be precisely foreseen. It is necessarily difficult to foresee as it is in the nature of an investigation. This is why the combination of items leads to the Employer's carrying some risks which in the excavation sections are carried by the Contractor.

Rule A1 adds the requirement that item descriptions for trial holes shall state the minimum plan area of the bottom of the hole or, where the work is undertaken to locate services, the maximum length.

The items for boreholes, both light cable percussion boreholes and rotary drilled boreholes, are also divided between number and depth. In both items, the nominal diameter is required to be stated in descriptions (rules A4 and A5). Rule A6 requires that rotary drilled boreholes which are continuations of light cable percussion boreholes should be identified separately. Rule D3 requires that the maximum depth used for classification of such rotary drilled boreholes should be measured from the Commencing Surface of the light cable percussion borehole through which they are drilled. This is because the drilling rate is affected by the total depth of the hole. The depth used for classification in the third division is the maximum depth of a group of holes.

The items for samples, site tests and observations, instrumental observations and laboratory tests are set down in the classification at B 4–7 * *. Most of the items are enumerated. The exceptions are permeability tests which are measured by time, and installations of pressure heads and inclinometers which are measured by length. Various additional descriptions are required to be given as shown in rules A8–A22. There are few other complications as demonstrated by the almost total absence of measurement, definition and coverage rules.

In CESMM2, items are now given for measurement of professional services in connection with ground investigation (B 8 * *). This work is only measured when the ground investigation contractor is expressly required to carry out analysis of records and results (rule M2). The items are deemed to include preparation and submission of reports (rule C4). This includes the provision of however many copies of reports are specified and transmission of them to whatever points are specified. It should be noted that rule C1 is a general coverage rule which states that the items for the ground investigation itself include preparation and submission of records and results. Again, this can be taken to include whatever is specified for provision and transmission of copies of records and results. When rule C1 is read alongside rules M2 and C4 it will be clear that there is a precise distinction between 'records and results' which are covered by the items B 1–7 * * and 'analysis' of those same records and results which is covered by the professional services items at B 8 * *.

*Schedule of changes in CESMM2*

1. Pumping test wells are not classified.
2. Ranges of nominal diameter for boreholes are deleted.
3. Arrangement of depth items for trial holes and boreholes are changed.
4. Samples, site tests and observations, instrumental observations and laboratory tests are generally expanded.
5. Items for professional services in the analysis of records and results are added.

| Number | Item description | Unit | Quantity | Rate | Amount £ | p |
|---|---|---|---|---|---|---|
| | GROUND INVESTIGATION. | | | | | |
| | TRIAL HOLES. | | | | | |
| B113 | Number in material other than rock maximum depth 2 - 3 m; minimum plan area at the bottom of the excavation 2 m2. | nr | 6 | | | |
| B130 | Depth in material other than rock; minimum plan area at the bottom of the excavation 2 m2. | m | 18 | | | |
| B150 | Depth supported. | m | 18 | | | |
| B160 | Depth backfilled with excavated material. | m | 18 | | | |
| B170 | Removal of obstructions. | h | 5 | | | |
| B180 | Pumping at a minimum extraction rate of 7000 litres per hour. | h | 12 | | | |
| | LIGHT CABLE PERCUSSION BOREHOLES. | | | | | |
| | Nominal diameter at base 150 mm. | | | | | |
| B210 | Number. | nr | 8 | | | |
| B231 | Depth in holes of maximum depth not exceeding 5 m. | m | 20 | | | |
| B232 | Depth in holes of maximum depth 5 - 10 m. | m | 10 | | | |
| B233 | Depth in holes of maximum depth 10 - 20 m. | m | 15 | | | |
| B234 | Depth in holes of maximum depth 20 - 30 m. | m | 50 | | | |
| B260 | Depth backfilled with excavated material. | m | 95 | | | |
| B270 | Chiselling to prove rock or to penetrate obstructions. | h | 10 | | | |
| | | | | PAGE TOTAL | | |

| Number | Item description | Unit | Quantity | Rate | Amount £ | p |
|---|---|---|---|---|---|---|
| | ROTARY DRILLED BOREHOLES. | | | | | |
| | Minimum core diameter 75 mm. | | | | | |
| B310 | Number; continuations of light cable percussion boreholes. | nr | 2 | | | |
| B343 | Depth with core recovery in holes of maximum depth 10 - 20 m. | m | 30 | | | |
| B360 | Depth backfilled with cement grout. | m | 10 | | | |
| B370 | Core boxes, core 3.0 m long. | nr | 4 | | | |
| | SAMPLES. | | | | | |
| B412 | Disturbed samples of soft material from the surface or from trial holes: minimum 5 kg; Class 3. | nr | 18 | | | |
| B421 | Open tube samples from boreholes. 100 mm diameter x 450 mm long, undisturbed sample; Class 1. | nr | 40 | | | |
| B422 | Disturbed samples from boreholes: minimum 5 kg, Class 3. | nr | 25 | | | |
| B423 | Groundwater samples from boreholes: minimum 1 litre. | nr | 8 | | | |
| | SITE TESTS AND OBSERVATIONS. | | | | | |
| B512 | Groundwater level; taken at commencement and end of work each day. | nr | 8 | | | |
| B513 | Standard penetration test; in light cable percussion boreholes. | nr | 25 | | | |
| B515 | Vane test in borehole. | nr | 5 | | | |
| | | | PAGE TOTAL | | | |

| Number | Item description | Unit | Quantity | Rate | Amount £ | p |
|---|---|---|---|---|---|---|
| | INSTRUMENTAL OBSERVATIONS. | | | | | |
| | Pressure head; weekly observations, protected with chestnut fencing. | | | | | |
| B611 | Standpipes. | m | 18 | | | |
| B613 | Install covers. | nr | 3 | | | |
| B614 | Standpipe readings. | nr | 9 | | | |
| | LABORATORY TESTS | | | | | |
| | Classification. | | | | | |
| B711 | Moisture content BS 1377 nr. 1a. | nr | 10 | | | |
| B712 | Atterberg limits BS 1377 nrs. 2a, 3, 4. | nr | 20 | | | |
| B714 | Particle size analysis by sieve BS 1377 nr. 7a. | nr | 8 | | | |
| B715 | Particle size analysis by pipette or hydrometer BS 1377 nrs. 7c, 7d. | nr | 4 | | | |
| | Chemical content. | | | | | |
| B722 | Sulphate BS 1377 nrs. 9, 10. | nr | 16 | | | |
| B723 | pH value BS 1377 nr. 11a. | nr | 8 | | | |
| | Consolidation. | | | | | |
| B741 | Oedometer cell BS 1377 nr. 17. | nr | 6 | | | |
| | Strength. | | | | | |
| B761 | Quick undrained triaxial BS 1377 nr. 21; set of three 38 mm diameter specimens. | nr | 20 | | | |
| | | | PAGE TOTAL | | | |

| Number | Item description | Unit | Quantity | Rate | Amount £ | p |
|---|---|---|---|---|---|---|
| | PROFESSIONAL SERVICES. | | | | | |
| B831 | Engineer or geologist graduate. | h | 5 | | | |
| B832 | Engineer or geologist chartered. | h | 15 | | | |
| B833 | Engineer or geologist principal or consultant. | h | 5 | | | |
| B840 | Visits to the Site. | nr | 2 | | | |
| | | | PAGE TOTAL | | | |

# *Class C: Geotechnical and other specialist processes*

Class C is a mixture of different types of work which have only their specialist nature in common. They are normally carried out by subcontractors. Each type of work is likely to generate its own Method-Related Charges—at least for bringing plant to the Site. Class C is considerably modified in the second edition of the CESMM.

The measurement of grouting (which excludes grouting from within tunnels, sewers, etc.) divides the work between that associated with forming the holes (C 1–4 * *) and that associated with the material and its injection (C 5 * *).

The holes may be formed either by drilling or by driving grout injection pipes. Separate drilling items are given for drilling through soft ground and for drilling through hard ground. Thus a single hole may require two items for drilling. There is only one item for driving grout injection pipes as they cannot be driven through hard material.

Drilling and driving items are classified in items C 1–3 1–5 * according to the zones of inclination shown in Fig. 12. Further itemization is required for holes of differing depths (measured from the Commencing Surface along the axis of each of the holes—rule M2) and for holes of differing diameters (which are stated—rule A1).

Classification C 4 * 0 lists features that are particular to the grout holes where the cost is independent of the length of drilling or driving. They are also subdivided according to the diameters of the holes (rule A1). The 'number of stages' measured will be the sum of the number required for each of the individual holes (rule M4). It is important to note the difference between drilling through a previously grouted stage in the course of descending stage grouting and drilling through a previously grouted hole not in the course of stage

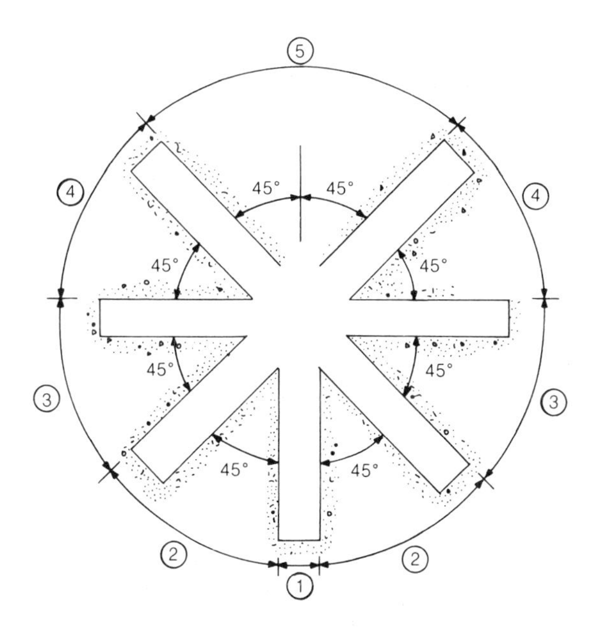


Fig. 12. Zones of inclination for grout hole drilling and driving given in the second division of class C, item codes C 1–3 1–5 *. Notice the precise boundaries for the zones. For example, drilling at 45° to the vertical upwards is in zone 4, drilling at 44° to the vertical upwards is in zone 5

grouting to extend it after the initial grout has set. The critical factor is the timing. Where descending stage grouting is being carried out in a normal sequence, redrilling usually proceeds within about twenty-four hours of grouting a stage. The hole can then be very easily and quickly redrilled to the previously grouted depth. If redrilling is instructed several days, weeks or months later, the operation is equivalent to drilling through concrete or rock. The CESMM reflects these cost differentials by identifying the number of stages at classification C 4 2 0. The resulting bill item(s) can include the cost of stage redrilling. Rule M3 requires holes which are drilled through set grout as a separate operation from stage grouting to be itemized separately.

The grout mix components can be varied during the course of a single injection. For this reason the CESMM provides separate items for the mass of materials supplied (C 5 1 *) and the mass mixed and injected (C 5 2 *). Both measurements exclude the mass of water used to mix the grout materials (rules M5 and M7). The broad categories of grout types at classification C 5 2 2–5 are sufficient to reflect the different labour and plant costs. The number of injections is measured at classification C 5 2 1 and is divided between those in ascending and those in descending stages (rule A3) as there is a considerable difference in cost between the two methods.

The measurement rules for diaphragm walls are simple compared with those for other excavation and concrete work. This acknowledges that construction of diaphragm walls is a specialist operation in which details of design affecting construction technique, such as guide walls and positioning of joints, are usually left to the Contractor. Note that excavation for diaphragm walls includes additional excavation for guide walls. If the Commencing Surface for the excavation is not the Original Surface, item descriptions must say so to comply with paragraph 5.21.

The CESMM does not require separate items to be given for trimming either the top or exposed side surfaces of diaphragm walls, for preparing protruding reinforcement for receipt of capping beams or for disposal of excavated material or indeed for any of the work referred to in rules C1–C4. It is consequently necessary for it to be made clear in sub-contract enquiries and quotations whether this work is to be covered by the sub-contract rates or whether the main Contractor intends to carry it out and allow for it in the main Contract rates.

The rules and terminology for ground anchorages in CESMM2 reflect the new British Standard code definitions and current international practice.

Soil is highly variable and anchorages are often installed in a mixture of sands and gravels or clays. It is therefore inappropriate to subdivide soil other than by stating that it does not contain hard material. Where anchorages are installed through soil and into rock it is the presence of the hard material which dictates the plant used for both the soft and the hard drilling. For this reason, if a hole includes any hard material it will all fall within the classification C 7 3–4 *.

The third division of classification identifies anchorages as being either temporary or permanent. Temporary anchorages are those installed for a known short period of time (usually less than two years) and, although they are not normally protected, in certain aggressive corrosive environments special protective measures may be required. Permanent anchorages have a service life of greater than two years and the tendons (wires, strands or bars) are normally protected from corrosion throughout their length. In rare circumstances, such as in low permeability concrete or rock for dam stressing, a permanent anchor may have no protection.

Rule A7 requires details of testing and pregrouting to be given when this work is required.

Vertical consolidation drains are measured at classification C 8 * *. Rule A8 requires separate items for each of the three common types of drains to be given. Rule M14 limits the measurement of predrilled holes to those which are expressly required to be predrilled through overlying material. It follows that an item for the depth of overlying material will only be measured where there is overlying material such as existing fill.

*Schedule of changes in CESMM2*

1. The classification table for grouting work contains new items for driving injection pipes for grout holes (C 3 * *) and for the number of stages of grouting (C 4 2 0).
2. Rules are added for the measurement of drilling through previously grouted holes and for the number of holes drilled.
3. Grout injections in ascending or descending stages are distinguished.
4. Rules for ground anchorages are expanded.
5. The classification table now includes predrilled holes for sand, band and wick drains and requires a maximum depth to be stated.

| Number | Item description | Unit | Quantity | Rate | Amount £ | p |
|---|---|---|---|---|---|---|
| | GEOTECHNICAL AND OTHER SPECIALIST PROCESSES. | | | | | |
| | Drilling for grout holes. | | | | | |
| | Diameter 50 mm. | | | | | |
| | Vertically downwards. | | | | | |
| C111 | In holes of depth: not exceeding 5 m. | m | 80 | | | |
| C112 | In holes of depth: 5 - 10 m. | m | 160 | | | |
| C113 | In holes of depth: 10 - 20 m. | m | 310 | | | |
| C114 | In holes of depth: 20 - 30 m. | m | 630 | | | |
| C116 | In holes of depth: 50 m. | m | 200 | | | |
| | Downwards at an angle $0^{o}$ - $45^{o}$ to the vertical. | | | | | |
| C123 | In holes of maximum depth: 10 - 20 m. | m | 620 | | | |
| C124 | In holes of maximum depth: 20 - 30 m. | m | 3300 | | | |
| | Grout holes diameter 50 mm. | | | | | |
| C410 | Number of holes. | nr | 234 | | | |
| C420 | Number of stages. | nr | 702 | | | |
| C430 | Single water pressure tests. | nr | 20 | | | |
| | | | PAGE TOTAL | | | |

| Number | Item description | Unit | Quantity | Rate | Amount £ | p |
|---|---|---|---|---|---|---|
| | Grout materials and injection. | | | | | |
| | Materials. | | | | | |
| C511 | Cement; to BS 12. | t | 154 | | | |
| C512 | Pulverized fuel ash; as Specification clause G420.7. | t | 290 | | | |
| C513 | Sand; to BS 1199. | t | 50 | | | |
| | Injection. | | | | | |
| C521.1 | Number of injections; in descending stages. | nr | 50 | | | |
| C521.2 | Number of injections; in ascending stages. | nr | 75 | | | |
| C522 | Neat cement grout. | t | 25 | | | |
| C523.1 | Cement and sand grout. | t | 200 | | | |
| C523.2 | Cement and pulverised fuel ash grout. | t | 269 | | | |
| C526 | Single packer settings. | nr | 50 | | | |
| | Diaphragm wall for anchor chamber as drawing 134/65 thickness 1.05 m. | | | | | |
| C613 | Excavation maximum depth 10 - 15 m. | m3 | 2400 | | | |
| C623 | Excavation in rock maximum depth 10 - 15 m. | m3 | 400 | | | |
| C640 | Concrete; with sulphate resisting cement strength 30 N/mm2. | m3 | 2730 | | | |
| C664 | High yield steel bar reinforcement to BS 4449 diameter 12 mm. | t | 40 | | | |
| C665 | High yield steel bar reinforcement to BS 4449 diameter 16 mm. | t | 66 | | | |
| | | | PAGE TOTAL | | | |

| Number | Item description | Unit | Quantity | Rate | Amount £ | p |
|---|---|---|---|---|---|---|
| | Diaphragm wall for anchor chamber as drawing 134/65 thickness 1.05 m. | | | | | |
| C666 | High yield steel bar reinforcement to BS 4449 diameter 20 mm. | t | 120 | | | |
| C670 | Waterproofed joints; as Specification clause C47. | sum | | | | |
| C680 | Guide walls. | m | 242 | | | |
| | Temporary ground anchorages; East Quay wall; horizontal load 27 tonnes/metre; water and grout testing, pregrouting and grouting as Specification clause W7/27. | | | | | |
| C711 | Number maximum depth 25 m. | nr | 7 | | | |
| C721 | Total length of tendons. | m | 140 | | | |
| | Sand drains. | | | | | |
| C811 | Number of drains cross sectional dimension not exceeding 100 mm. | nr | 50 | | | |
| C851 | Depth of drains of maximum depth: 10 - 15 m cross sectional dimension not exceeding 100 mm; type C2 granular material. | m | 650 | | | |
| | | | | PAGE TOTAL | | |

# *Class D: Demolition and site clearance*

Demolition and site clearance are described and itemized in fairly brief terms using the CESMM. The rules of measurement are based on the assumption that demolition sub-contractors need to inspect the buildings and structures to be demolished before pricing and that they need to do so however detailed the Bill of Quantities. Similarly, site clearance costs depend on the characteristics of the terrain, what machines can move about the site, whether the vegetation is dense or non-existent, whether or not walls and fences will yield saleable material and so on. As inspection is essential and variations are infrequent, the bill can be a simple check list of items for trees, stumps and structures. The clearance from the Site of all smaller things can be given as a sum, but to avoid change for change's sake, the CESMM gives it as a price per hectare of area to be cleared. Rule A2 says that the tenderer may assume that the area given for general clearance is the whole Site area unless the item description identifies only a part of this area.

Buildings and other structures have to be identified in item descriptions to comply with rule A4. This ensures that the price is related to the particular ease or difficulty of demolition of each building and to the particular value of the materials it contains. The identification should be either in terms of a clear name or number of the building or a reference to identification numbers or letters given on a plan of the Site and included in the tender drawings.

The former requirement to classify clearance areas by type of land has been deleted in CESMM2. Although there is a considerable difference in cost between clearing, for example, urban land and open moorland, tendering contractors are unlikely to remain unaware of what they are pricing in the absence of a specific description.

The predominant material and the range of approximate volume occupied must be given for buildings and other structures. This is really only a guide for comparison of prices from one bill to another, as the price is for the buildings named and identified in the item descriptions. All structures and buildings, however small, have to be identified. Identical structures can be grouped together so that demolition of a row of fifty beach huts would not need fifty bill items.

Tenderers will assume that all materials derived from site clearance and demolition will become the Contractor's property unless item descriptions state otherwise (rule A1). This also means that they must remove the materials from the Site as they are required to do by clause 33. Rule A1 also means that separate items must be given for the demolition of things which are to remain the property of the Employer.

There is a potential problem is determining where demolition and site clearance end and excavation begins. The CESMM clarifies this by establishing in the includes and excludes list for class D that, except for tree roots, the dividing line is the Original Surface of the Site. This is defined in paragraph 1.10 as the surface of the ground before any work is done, work being defined in paragraph 1.5 as work to be undertaken by the Contractor under the Contract. This does not remove all possible ambiguities. There might be a slab of concrete which is two thirds below the surrounding ground level and one third above it. Strictly the protruding part should be broken away as 'demolition' and the buried part left to be dug up as 'excavation of artificial hard material: concrete'. Bill compilers should be aware of this difficulty. They should make clear by means of additional description any intention that more work than removing everything which is sticking up above the surrounding ground is to be regarded as work in class D.

Similarly, when working out the approximate volume occupied by buildings and structures, basements should be excluded. Volume occupied includes the volume of the space enclosed by a building, and is not just the volume of material in its walls, floors and roof. This criterion also cannot be applied too literally. The volume occupied by an electricity transmission pylon or a cast iron gasometer frame is not a very helpful measurement, but as such structures are identified no problem is created.

Rule D1 establishes that the rules for itemization of demolition and site clearance are different from those for other classes in one respect. This is that the components of work referred to in items D 2–6 * * are those particular articles, buildings or separate structures which are to have separate items to distinguish them from the general site clearance which is covered in items D 1 0 0. Thus the lists of size ranges for trees, stumps and pipelines do not start at zero. Separate items are not given for trees of girth 500 mm or less, stumps of diameter 150 mm or less or pipelines of nominal bore 100 mm or less. The cost of their demolition and clearance is to be included in the rate for general clearance. This means that hedges are not measured separately, but that if a shrub or bush within or separated from a hedge has a stem of girth exceeding 500 mm one metre above ground level, it would be added to the number of trees measured in item D 2 1 0.

*Schedule of changes in CESMM2*

1. Categories of land to be cleared are eliminated.

| Number | Item description | Unit | Quantity | Rate | Amount £ | p |
|---|---|---|---|---|---|---|
| | PART 5. DEMOLITION AND SITE CLEARANCE. | | | | | |
| D100 | General clearance. | ha | 27 | | | |
| D220 | Trees girth 1 - 2 m; holes backfilled with excavated material. | nr | 39 | | | |
| D250 | Trees girth exceeding 5 m. | nr | 30 | | | |
| D320 | Stumps diameter 500 mm - 1 m. | nr | 21 | | | |
| D414 | Buildings brickwork volume 250 - 500 m3; farmhouse at Farley Court. | sum | | | | |
| D435 | Buildings masonry volume 500 - 1000 m3; barn at Farley Court. | sum | | | | |
| D545 | Other structures metal volume 500 - 1000 m3; screening plant to remain the property of the Employer. | sum | | | | |
| D620 | Pipelines nominal bore 300 - 500 mm. | m | 110 | | | |
| D900 | Pump removed, cleaned and delivered to Employer's depot at Church Stretton. | nr | 2 | | | |
| | | | | PAGE TOTAL | | |

# *Class E: Earthworks*

Earthworks occur in nearly all civil engineering contracts. Their relatively unpredictable extent and the susceptibility of their cost to weather and water conditions make estimating for them an uncertain science. The fortunes of contractors often depend on the turn of the wheel of earthmoving costs.

The methods of measurement for earthworks reflect this uncertainty but try to compensate for it by providing items which can be priced to model those factors which most influence earthmoving costs. Only in this way can the Employer's confidence in paying only for what is actually done on his behalf be maintained and can the Contractor continue to regard earthmoving as a normal business activity and not as a grotesque gamble. The methods of measuring earthworks were substantially overhauled in preparing the second edition of the CESMM. As a result, a small number of quite significant changes were made. Each of these is explained in this chapter.

A new rule M1 governs the calculation of quantities for earthworks as a whole. The first sentence reinforces paragraph 5.18 of the general rules. The second sentence of M1 is necessary as it is not normally practical to measure the volumes of items such as topsoil, rock or artificial materials from drawings. They are normally recorded from site surveys and observations.

The first five features of the first division of class E cover excavation, the next two filling and the last landscaping. The main principle of the excavation items is that they should cover excavation of one type (dredging, cuttings, foundations or general excavation) in one type of material (topsoil, rock, other natural material or artificial hard material). In addition the items for foundations and general excavation are classified according to the range of depth below the

Commencing Surface in which the maximum depth of the excavated hole occurs.

The depths stated for general and foundation excavation are governed by paragraph 5.21, the definitions in paragraphs 1.10–1.13 and rule M5. These provisions of the CESMM establish a principle to be adopted for separation of excavation work into items where excavation may or must be carried out in several stages. If an item description read simply 'excavation for foundations', this would mean that the depth of excavation was to be taken from the surface of the ground before the Contract work began to the surface of final excavation immediately above which new permanent work, such as blinding concrete, was to be placed. If it is required to limit the depth of work included in any one excavation item to a stage of excavation, this must be done by stating either the Commencing Surface at which the stage of excavation work in that item is to begin or the Excavated Surface for that item. The maximum difference in level between the two surfaces is the depth of excavation for that item. In accordance with paragraph 5.21 and rule A4 of class E, a statement of the Commencing Surface can be omitted where it is at the original ground level before work included in the Contract, and definition of the Excavated Surface can be omitted where this is the final level required to accommodate the permanent work or to complete the excavation included in the Contract. An example of this rule is shown in Fig. 7.

Excavation should not be divided into stages in the items in the bill indiscriminately as this could lead to unnecessarily complicated billing. Items E 3 1 1 to E 3 2 4.2 and E 2 2 0.1 and E 2 2 0.2 in the example bill illustrate the procedure in use. In the former case stripping of topsoil is shown as a stage of excavation, and in the latter case the excavation of the last 0·25 m in the depth of a cutting is shown as a stage of excavation. Use of this latter procedure would be a requirement of the CESMM (rule M5) if the Specification required a layer of material to be left for protection of the formation after main cutting excavation had been carried out.

Items E 4 1 5 to E 4 3 5 of the example bill demonstrate the more usual procedure in which intermediate surfaces are not defined. Here, to exaggerate the point, all three different materials are classed as 'maximum depth 2–5 m'. This means only that the materials occur within the depth of excavations whose maximum depth is between 2 m and 5 m. It would be reasonable to assume that the topsoil is at the surface. The rock and the soft material could occur anywhere within the depth of the hole, in alternating layers or in lumps. Note that no material is identified in the description of item E 4 2 5. This illustrates the application of definition rule D1. Where no material is stated, it is assumed to be ordinary soft material.

Paragraphs 1.12 and 1.13 as revised in CESMM2 when read with paragraph 5.21 have the effect that separate volumes of different material within one hole do not need to be measured with Commencing Surface and Excavated Surface stated separately. The depth stated in the third division for excavation of foundations and general excavation is essentially the depth of the hole to be dug, not the thickness of any bands of different material within the hole. This is to ensure that items and prices reflect the choice of plant and method appropriate to excavating within a particular hole. Paragraphs 1.12 and 1.13 were changed in CESMM2 in order to make this more clear than it was in the first edition.

It is permissible, in special circumstances, to distinguish between two materials which occur in a known sequence in the depth of a hole. For example, it might be necessary to excavate boulder clay from the Original Surface to the top surface of a massive rock formation and then remove 4 m depth of rock to form a key. It would be helpful, and encourage good pricing, if this were billed in two items, the first described as '. . . excavation maximum depth 10–15 m; Excavated Surface top of rock' and the second as '. . . excavation in rock maximum depth 2–5 m; Commencing Surface top of rock'.

It should be noticed that the four definitions of surface do not refer to levels. This is so that surfaces can be identified simply using terms like 'top of rock' or 'underside of road base'. If they were to mention 'level' it might be assumed that ordnance datum levels are required, which is not so. This would be too complicated for irregular surfaces and surfaces whose precise levels are unknown. The same terms are used for Commencing Surfaces for work such as boring and drilling where, because boring occasionally takes place from a vertical Commencing Surface, it would be misleading to refer to a level.

A common use of this procedure for referring to intermediate surfaces is when it is necessary to clarify where general excavation or excavation of cuttings stops and excavation of foundations begins. Rule A3 refers to this problem. If, for example, a retaining wall is to be constructed along the side of a cutting there could be no rule saying which volume of excavation is forming the cutting and which volume is accommodating the retaining wall and its foundations. Whatever convention is used it is not certain to reflect the methods of excavation eventually adopted by the Contractor. Whatever boundary betwen the volumes measured in each of the two items is adopted by the compiler of the bill, it will be an Excavated Surface in one item which will be the Commencing Surface for the other. In many cases it will be easiest and clearest to show the boundary as 'payment line A–A' on the Drawing and to refer to this, with the Drawing number, as the definition of the surfaces in the item descriptions.

Rule M6 provides the convention for measurement of the volume of excavation to accommodate foundations and the structures above them. 'Any part of the structure or foundation' means what it says; the part does not have to be a structural part. If, for example, a concrete tank has a perimeter drain, the excavation should be measured to include the drain and its bed or surround. The volume to be measured for such excavation is the volume of the hole into which it would be notionally possible just to drop the complete structure vertically to its actual depth. Figure 13 shows this for two particularly irregularly shaped structures. This rule may produce some measurement of volumes of excavation not occupied by structures, as the figure shows, but no further volumes will be measured to acknowledge any necessity for working space around the structures. Tenderers have to allow in their excavation rates or associated Method-Related Charges for the cost of excavating and backfilling any volume they judge necessary for working space, and for any ancillary costs such as dealing with obstructions in working space. The space necessary to accommodate Temporary Works generally, such as formwork and supports to excavations, is regarded as working space and is not measured.

In CESMM2 disposal of excavated material is measured separately from excavation itself in order to enable the balance between excavation and filling to be calculated simply. Material will be

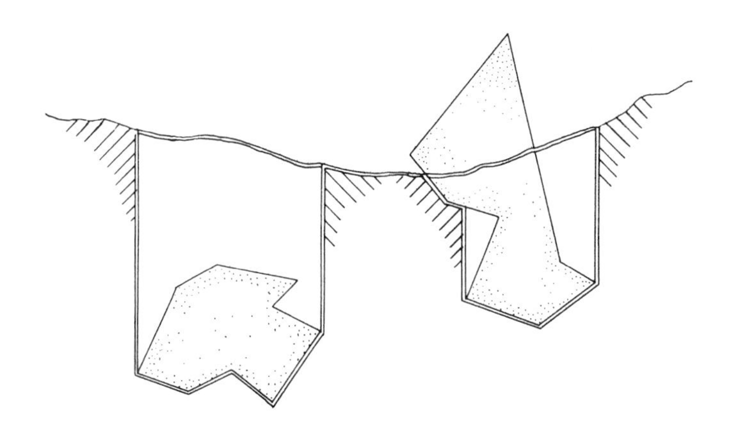

Fig. 13. These ridiculously shaped structures, which are to be constructed after excavation for them has been carried out, demonstrate the application of rules M6 and M16. The volume measured for excavation is the volume bounded by the double line. The volume measured for backfilling, if required, would be the part of this volume not occupied by the finished structure

classed as for disposal either because it is surplus to filling requirements or because it is unsuitable for use as filling. The CESMM does not require this distinction to be made in the Bill of Quantities as it is not cost significant. The difference between the volumes of excavation and of disposal of excavated material is equal to the volume of excavated material re-used on Site. Rule D4 states that disposal is deemed to be off the Site unless otherwise stated. If disposal is on the Site, the location is to be described in accordance with rule A8. Where the volume of the filling material measured exceeds the volume of the excavated material which is to be re-used, a volume of imported material is measured. Rule A9 requires the type of imported material to be identified.

For CESMM2 the arrangements for measurement of disposal of excavated material have been changed in other classes. It is consistent in that work in any other class which involves excavation and which might generate surplus excavated material is dealt with in the same way. In diaphragm walling, pipe laying in trenches, piling, etc., all the items involving excavation are deemed to include disposal of surplus excavated material. If the Contractor chooses to dispose of it on the Site by using it as filling material, and is permitted to do so by the Specification, the effect of doing so is ignored in calculating the quantities in the earthworks items. The only exception to this is that the use of excavated material derived from tunnelling as fill material is taken into account. This is the effect of rules M19 of class E and A4 of class T. In special circumstances, such as where a contract includes excavation for considerable lengths of very large diameter pipes, an exception can be made by including an appropriate statement amending the CESMM in the Preamble to the Bill of Quantities.

Excavation from borrow pits is covered by rules M9, D3, C2 and A5 of class E. The items for such excavation are deemed to include removal and replacement of overburden and any unsuitable material. The volume measured is the volume extracted from the borrow pit which is actually used for filling. By virtue of rule M9, this volume is measured as the filled volume, unlike other excavation items where the volume measured is the volume of the void formed by the excavation.

Where there are different materials which have either a very different cost of excavation or of disposal within a classification given in the CESMM additional description should be given and the item subdivided in accordance with paragraph 5.10. However, distinctions between different soft excavated materials and different hard excavated materials or different types of rock are very difficult to make clearly. Having made the distinction in one part of the Bill of Quantities an argument could be mounted that similar distinctions should have been made in other parts of the Bill of Quantities. Consequently, applications of paragraph 5.10 for this purpose should be considered very carefully and either kept to a minimum or associated with extremely careful wording.

The CESMM does not require running sand to be identified separately as a material to be excavated. This is because dealing with all aspects of groundwater is regarded in the Conditions of Contract as a contractor's risk. The phenomenon of running sand is a function of groundwater movements and pressures. If the presence of running sand 'could not reasonably have been foreseen by an experienced contractor' as stated in clause 12 of the Conditions of Contract, the cost of dealing with it can be recovered by the Contractor. Otherwise he will be deemed to have made an allowance for the risk in his rates.

The effect of rule M3 is that where a contractor chooses to carry out work classed as dredging in the Bill of Quantities by other means, such as by excavating within cofferdams, the work will still be classed as dredging and the original bill rates will apply. This rule exists because of the difficulty of forecasting when the Contractor will employ dredging and when he will employ other means of excavating below water. The bill compiler must consider which means is most likely to be used and measure the work accordingly.

Rule M4 states that dredging normally should be measured from soundings. Where this method of measurement may not be possible, the Preamble to the Bill of Quantities should state that hopper measurements may be used and give the circumstances in which they will be used. The normal method of calculating the volume of excavation by soundings is to take soundings before work begins and after work has been completed in order to calculate the volume of the void formed by the excavation. There is what might appear to be a small change in CESMM2 which affects the measurement of dredging. The first division entry in the classification table E 1 * * which is now called 'Excavation by dredging' was formerly called 'Dredging'. The change was made so that dredging should be regarded quite clearly as a type of excavation, not as something different. The significance of the apparently unimportant change is that all the rules which refer to excavation now uniquivocally apply to dredging. This means, for example, that where excavation ancillaries have to be measured for excavation in the dry they also have to be measured for dredging.

Dredging to remove silt is classed as an excavation ancillary. In the first edition of the CESMM this work was only measured if it was expressly required to be carried out during the maintenance period for the contract as a whole. The new rule M14 in CESMM2 requires this work to be measured if removal of silt is expressly required at any time after the dredging has once been completed and the Final Surface reached. This change was made so that the Contractor should not carry the risk of silt accumulating in the period between completion of the dredging and the end of the maintenance period for the whole contract.

Rule M13 provides that double handling of excavated material shall be measured only to the extent that it is expressly required. Consequently, no stockpiling of material will be paid for at the rates against items E 5 4 * unless stockpiling is expressly required. If the Contractor stockpiles without being instructed to, however sensible or difficult to avoid stockpiling might be, he will not be paid separately for it. The items for topsoil demonstrate this. 'Topsoil' appears as an excavation item and 'excavated topsoil' appears as a filling item. Somewhere between the two the Contractor will normally stockpile topsoil and consequently double handle it. If he is clever at programming he may be able to do some topsoiling without double handling. However, the Employer would not mind if he kept the topsoil in parked scrapers until he was ready to spread it and certainly does not expressly require topsoil to be double handled. Care should be taken that the wording of specification clauses and notes on drawings referring to the method of forming or location of temporary stockpiles cannot be wrongly construed as meaning that double handling is expressly required.

The CESMM requires excavation of rock and artificial hard materials to be measured separately. Rule M8 qualifies this to the extent that an isolated volume of a hard material is not measured if its volume exceeds 1 $m^3$ or 0·25 $m^3$ in the case of an excavation which does not exceed 2 m in width. This rule recognizes that smaller pockets of hard materials do not significantly affect the cost of normal excavation unless they occur within narrow workings.

Rules A2 and M7 relate to the new arrangements for dealing with work affected by water adopted in CESMM2 and expressed in the revised paragraph 5.20. Rule A2 requires that excavation below bodies of open water such as a river, streams, canals, lakes, reservoirs or estuaries should be given in separate items for each body of water. Rule M7 provides the rule for calculating the quantity of excavation identified as being below water. In simple terms, it is the volume which is below the plan area of the body of water. If the body of water fluctuates in size or boundary owing to tidal or any other effects, it is the larger area which is taken. Figure 14 illustrates this point. The words of paragraph 5.20 and rule M7 are carefully chosen

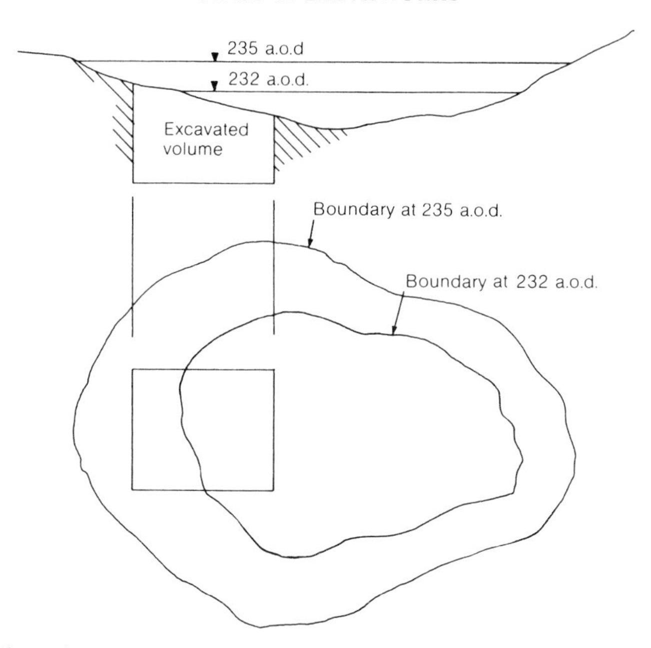


Fig. 14. The volume measured for excavation below water is that which is below water at the higher anticipated level

to prevent arguments developing about the difference between anticipated and actual levels or boundaries of bodies of water. Measurement of volume is unequivocally based on the anticipated levels and boundaries of the water, not what actually happens. The only qualification to this would be that the occurrence of totally unanticipated fluctuations in water level could be grounds for additional payment under clause 12 of the Conditions of Contract.

Reference to an obligation of the Contractor to keep excavations free of water has been deleted from the CESMM in this class and elsewhere. This is because the Contractor really has no such obligation. He is obliged to construct the Permanent Works in accordance with the Contract, dealing how he sees fit with any groundwater, rainwater or open water that he may encounter. He certainly is not obliged to remove the water if he can find an easier or more economic way of constructing the works.

The measurement of trimming and preparation of excavated and filled surfaces is changed in CESMM2. The arrangements are somewhat simpler than in the first edition. Surfaces left after excavation or filling has been carried out are measured and items given for their area. If the excavated or filled surface is not to carry any Permanent Works, the Contractor will have to trim it and an item will be measured for trimming. If the excavated or filled surface is to carry Permanent Works—except surfaces which are to be filled or landscaped and surfaces which are also measured for formwork, the Contractor will have to prepare the surface to receive them and an item will be measured for preparation. Preparation of surfaces to receive geotextiles is measured becasue geotextiles are not classed as filling. Preparation to receive a capping layer is also measured if the capping layer is classed as roadworks, not as filling.

In accordance with rules A7 and A14, both trimming and preparation items must have appropriate additional description when they are sloped at an angle exceeding 10°.

A significant change in CESMM2 is that the words 'whether trimming is expressly required or not' are included in rules M10 and M22 and similar words affecting preparation are included in rules M11 and M23. Such rules appear nowhere else in the CESMM. They are only necessary because the item descriptions concerned refer to an operation which the Contractor may or may not do, they are not recognizable components of the physically completed work. The words have been added because there were some misunderstandings about when trimming and preparation were measured using the rules of the first edition of the CESMM. The rules are now different and clearer. Another difference is that trimming is measured to horizontal surfaces.

The filling and compaction items generated by the CESMM distinguish filling to structures, forming embankments, filling in self-contained layers to stated depth or thickness and general filling. The former category of pitching is deleted from CESMM2. Filling material is also distinguished. Rule A10 requires compaction details to be stated if they are not uniform for each identified material. The CESMM rules always provide two items for an operation involving excavation, filling and compaction. Excavation rates cannot include filling and compaction. The CESMM does not give a definition of the distinction between embankments and filling to structures. In order to reflect cost differences properly, filling which may be termed 'embanking' when it is associated with structures should be measured as 'filling to structures'. Rule M16 refers to rule M6. These two rules, read with rule C1, mean that filling in order to backfill excavated working space is not measured.

Rule M19 gives the main convention for calculating quantities for filling and compaction items. It is derived from the general rule in paragraph 5.18 which is restated in rule M1 that quantities are measured net and that bulking and shrinkage of materials which are moved are ignored. The volumes calculated for imported filling and for disposal of surplus excavated material can be unrealistic as a result of this convention. Any bulking means either that less imported material is actually needed than is measured or that less surplus disposed of is measured than is actually removed. Similarly any shrinkage means either that more imported material is actually needed than is measured or that more surplus disposed of is measured than is actually removed. Rule M18 provides that the first 75 mm of settlement of the bottom of an embankment or of penetration into underlying material shall not be measured.

Despite the convention given in rule M18, there may be difficulty in determining the quantities involved in a cut and fill operation. Consider the general case shown in Fig. 15. The material excavated from a hole can be for three destinations

| | |
|---|---|
| quantity *a* | excavation for disposal |
| quantity *b* | excavation for re-use |
| quantity *c* | excavation for re-use after double handling |

Similarly the material in an embankment can come from three sources

| | |
|---|---|
| quantity *d* | imported fill |
| quantity *e* | excavated fill |
| quantity *f* | excavated double handled fill |

Quantities which can conveniently be physically measured are the volumes of

| | |
|---|---|
| quantity *A* | the hole |
| quantity *B* | the hole dug in the stockpile |
| quantity *C* | the embankment |

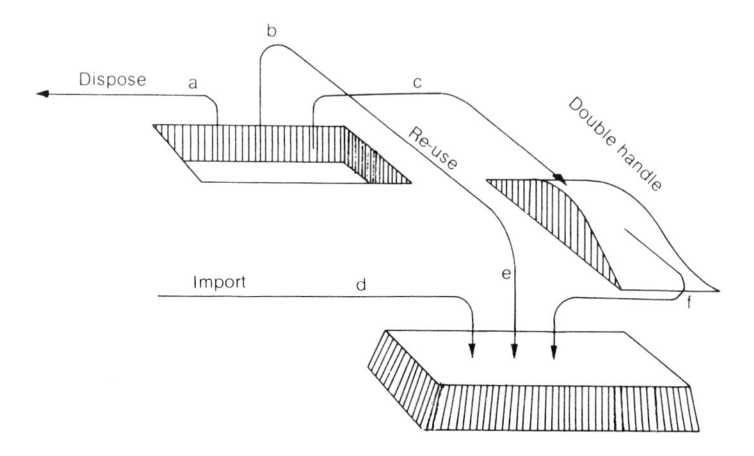


Fig. 15. The general case of the cut and fill operation. When a composite earth-moving operation such as that represented here is to be measured, care must be taken to ensure that enough quantities are measured in the field to enable the quantities against the bill items to be calculated

If no bulking or shrinkage is assumed, these nine quantities are related by the equations

$$
\begin{aligned}
A &= a + b + c \\
b &= e \\
C &= d + e + f \\
B &= f
\end{aligned}
$$

The quantities stated in the Bill of Quantities are

| | |
|---|---|
| Excavation E 4 2 * | $= a + b + c = A$ |
| Disposal of excavated material E 5 3 2 | $= a$ |
| Double handling of excavated material E 5 4 2 | $= f = B$ |
| Filling with excavated material E 6 3 3 | $= e + f = C - d$ |
| Filling with imported material E 6 3 5 | $= d$ |

These relationships show that it is not necessary to measure $b$ and $e$ to establish the quantities appearing in the bill. They also show that if there are volumes of disposal $a$ or import $d$ they may both have to be assessed by a secondary method such as a vehicle check or an apportionment at the point of excavation. In the general case there are four equations with six unknowns. This means that two measurements must be made by secondary means to make the equations soluble and the quantities calculable. Any two unknowns can be chosen; they should be those which are easiest to measure with acceptable accuracy.

If only one of the three operations disposal, double handling and import are present, the quantities are calculable without secondary measurement.

This general case is by no means the worst case. The difficulty is compounded when the excavated material is disposed of to more than one location and when the hole contains differently classed materials. Yet another dimension of confusion arises when several holes supply several stockpiles and several embankments and when several embankments are supplied from several holes and several stockpiles. A method of measurement cannot provide rules which deal with all possible combinations of this problem. It can only be said that it must be considered where the problem is likely to occur and the solution to be adopted must be communicated to tenderers via the Preamble to the Bill of Quantities.

Rule D7 has the effect that the Contractor may use excavated rock as filling wherever he chooses and the Specification permits, but that he will only be paid at the rates in the bill for filling using excavated rock where rock is expressly required to be used within filled volumes. Such use will normally be at stated locations within filled volumes, not spread indiscriminately throughout an embankment.

Rule M17 has the effect that where a Contractor has constructed a temporary road within a volume which is subsequently to be filled then, provided that the Engineer approves the road as being of suitable materials and construction to be included in the Permanent Works, the Contractor can include the volume of the road in the volume measured for filling.

The items for filling to stated depth or thickness (E 6 4 *) are used extensively as they include a wide variety of operations such as pitching, beaching, providing a drainage blanket, topsoiling or laying a filter medium. In the description of items the fact that filling is to a stated depth or thickness should take precedence over whether it is to structures or embankments. 'Stated thickness' means that the thickness must be stated and, consequently, that separate items must be given for different thicknesses.

Laying of geotextiles within or above earthworks is included in the CESMM2 as an additional item in filling ancillaries E 7 3 0. Rule A15 requires the material comprising the geotextile to be identified. The area measured for geotextiles would normally include any specified laps in accordance with paragraph 5.18 as there is no measurement rule to the contrary.

Some further comments on the measurement of trimming and preparation of surfaces may be helpful. In effect, trimming is only measured to surfaces which will remain visible when the Permanent Works have been completed. Preparation is measured for surfaces which are to have concrete cast against them. Preparation and trimming are not measured for temporary surfaces between layers of fill or between excavation and filling. Preparation and trimming are also not measured if any work classed as landscaping is to follow as such work is usually carried out by a landscaping sub-contractor as an integral part of his own work. Rule C4 establishes that the landscaping items themselves are deemed to cover trimming and preparation of surfaces.

*Schedule of changes in CESMM2*

1. Borrow pit excavation is measured separately and is to include dealing with overburden.
2. Disposal of excavated material is measured separately.
3. Trimming and preparation rules are substantially revised.
4. Filling to form temporary roads is included in the volume of fill measured.
5. Items for geotextiles are included in the classification tables.
6. Measurement rules for calculating volumes of fill are revised for materials arising from other classes of work.
7. Renumbering of items generally.
8. New arrangements for identifying work affected by water.
9. Dredging to remove silt accumulation is measured if carried out before completion.

| Number | Item description | Unit | Quantity | Rate | Amount | |
|---|---|---|---|---|---|---|
| | | | | | £ | p |
| | EARTHWORKS. | | | | | |
| E120 | Excavation by dredging; Mersey estuary adjacent to Sandon Dock. | m3 | 10971 | | | |
| E311 | Excavation for foundations topsoil maximum depth: not exceeding 0.25 m. | m3 | 37 | | | |
| E324 | Excavation for foundations maximum depth: 1 - 2 m. | m3 | 170 | | | |
| E324.1 | Excavation for foundations maximum depth: 1 - 2 m; around pile shafts. | m3 | 23 | | | |
| E324.2 | Excavation for foundations maximum depth: 1 - 2 m; Commencing Surface underside of blinding of outfall. | m3 | 15 | | | |
| | General excavation. | | | | | |
| E415 | Topsoil maximum depth: 2 - 5 m. | m3 | 153 | | | |
| E425 | Maximum depth: 2 - 5 m | m3 | 17350 | | | |
| E435 | Rock maximum depth: 2 - 5 m. | m3 | 750 | | | |
| E444 | Mass concrete exposed at the Commencing Surface maximum depth: 1 - 2 m. | m3 | 53 | | | |
| E454 | Reinforced concrete not exposed at the Commencing Surface maximum depth: 1 - 2 m | m3 | 112 | | | |
| | Excavation ancillaries. | | | | | |
| E512.1 | Trimming of excavated surfaces. | m2 | 7500 | | | |
| E512.2 | Trimming of surfaces; excavated by dredging. | m2 | 5271 | | | |
| E522 | Preparation of excavated surfaces. | m2 | 15763 | | | |
| | | | | PAGE TOTAL | | |

| Number | Item description | Unit | Quantity | Rate | Amount £ | p |
|---|---|---|---|---|---|---|
| | EARTHWORKS. | | | | | |
| | Excavation ancillaries. | | | | | |
| E523 | Preparation of excavated rock surfaces. | m2 | 576 | | | |
| E524 | Preparation of excavated surfaces; mass concrete. | m2 | 110 | | | |
| E532.1 | Disposal of excavated material. | m3 | 2700 | | | |
| E532.2 | Disposal of material excavated by dredging. | m3 | 10971 | | | |
| E542 | Double handling of excavated material. | m3 | 10350 | | | |
| E560 | Excavation of material below the Final Surface and replacement with grade B granular material. | m3 | 1000 | | | |
| E570 | Timber supports left in. | m2 | 300 | | | |
| E580 | Metal supports left in. | m2 | 500 | | | |
| E613 | Filling to structures. | m3 | 527 | | | |
| E624 | Filling to embankments, selected excavated material other than topsoil or rock. | m3 | 100716 | | | |
| E712 | Filling ancillaries trimming of filled surfaces; inclined at an angle of $10^{\circ}$ - $45^{\circ}$ to the horizontal. | m2 | 1054 | | | |
| E722 | Filling ancillaries preparation of filled surfaces. | m2 | 357 | | | |
| | | | PAGE TOTAL | | | |

| Number | Item description | Unit | Quantity | Rate | Amount | |
|---|---|---|---|---|---|---|
| | | | | | £ | p |
| | PART 27. ROADS. | | | | | |
| | EARTHWORKS. | | | | | |
| E210 | Excavation for cuttings; topsoil. | m3 | 584 | | | |
| E220.1 | Excavation for cuttings; Excavated Surface 0.25 m above Final Surface. | m3 | 10587 | | | |
| E220.2 | Excavation for cuttings; Commencing Surface 0.25 m above Final Surface. | m3 | 680 | | | |
| E522 | Excavation ancillaries preparation of excavation surfaces. | m2 | 5110 | | | |
| E532 | Excavation ancillaries disposal of excavated material; to on Site spoil heap at location A on drawing 7/47. | m3 | 584 | | | |
| E624 | Filling embankments selected excavated material other than topsoil or rock. | m3 | 5834 | | | |
| E712 | Filling ancillaries preparation of filled surfaces. | m2 | 2787 | | | |
| E730 | Filling ancillaries geotextiles; melded fibre mat grade G2 laid upon surfaces inclined at an angle of $10^{o}$ - $45^{o}$ to the horizontal. | m2 | 1250 | | | |
| | | | | PAGE TOTAL | | |

| Number | Item description | Unit | Quantity | Rate | Amount £ | Amount p |
|---|---|---|---|---|---|---|
| | PART 28. SITEWORKS. | | | | | |
| | EARTHWORKS. | | | | | |
| E411 | General excavation topsoil maximum depth: not exceeding 0.25 m. | m3 | 3482 | | | |
| E424 | General excavation maximum depth: 1 - 2 m. | m3 | 25234 | | | |
| E532 | Disposal of excavated material. | m3 | 9924 | | | |
| E624 | Filling embankments selected excavated material other than topsoil or rock. | m3 | 15310 | | | |
| E641 | Filling thickness 150 mm excavated topsoil. | m2 | 23213 | | | |
| E642 | Filling thickness 150 mm imported topsoil. | m2 | 1573 | | | |
| E711 | Filling ancillaries trimming of filled surfaces, topsoil. | m2 | 24786 | | | |
| E810 | Landscaping turfing. | m2 | 5700 | | | |
| E830 | Landscaping grass seeding upon a surface inclined at an angle to the horizontal exceeding $10^{\circ}$. | m2 | 2050 | | | |
| E860.1 | Landscaping oak trees exceeding 5 m high. | nr | 110 | | | |
| E860.2 | Landscaping sycamore trees exceeding 5 m high. | nr | 120 | | | |
| | | | | PAGE TOTAL | | |

# *Class F: In situ concrete*

The rules in class F of the CESMM generate items for the provision of concrete and separate items for placing concrete. A small group of items is given in the Bill of Quantities for the provision of concrete: one item for each mix to be used in work classed as in situ concrete. These items are followed by a larger group for placing concrete. The items in this group distinguish different types of concrete (mass, reinforced and prestressed) and distinguish concrete placed in different locations and in different structural elements of different sizes. These items do not distinguish different concrete mixes.

The revisions in the second edition of the CESMM have little effect on the measurement of in situ concrete. The only significant change is the modernized description of concrete mixes in the provision items in terms of BS 5328 replacing CP 110.

The object of giving separate items for the provision and placing of concrete is to simplify estimating and to make valuation of variations more realistic in relation to costs. Estimators do not have to extract the volumes of various mixes of concrete from a large number of items. The rates for placing concrete need only allow for labour and small tools applied to placing and curing concrete. Any major plant and labour involved in batching and transport of concrete can be covered by Method-Related Charges. Should there be any variations which necessitate changes in rates, the assessment of the rates will be made in terms of either a changed mix or changed placing conditions. The rate for one or other type of item is changed without conflict with other rates.

The item descriptions for provision of concrete use the terminology of BS 5328,* particularly the form of specification of concrete

* *Methods of specifying concrete, including ready mixed concrete.* British Standards Institution, London, 1981, BS 5328.

mixes which that document gives in paragraphs 5 and 12. The proportions of a designed mix are selected by the Contractor; those of a prescribed mix are selected by the Engineer.

Ordinary structural concrete (items F 1–3 * *) may only be made with cement to BS 12 or BS 146 or BS 4027 for sulphate-resisting cement. The possible grades are C7.5 and C10 (F 1 1 *) and C15–C30 (F 1 3–6 *). The permitted maximum size of aggregate is either 10, 14, 20 or 40 mm as shown in the third division of class F. Additional information required to specify an ordinary prescribed concrete mix in accordance with BS 5328 fully comprises the permitted types of aggregate in terms of BS 882 and 1047. A statement about workability may also be given but is not a requirement. To comply with rule A1 in class F, item descriptions for ordinary prescribed mixes must therefore state the permitted types of aggregate in addition to the information drawn from the classification table itself.

Designed mixes require a more comprehensive specification to comply with BS 5328 and, consequently, with rule A1 of class F. In addition to the information drawn from the classification table, the description of a designed mix (F 2 * *) must state the permitted size of aggregate, the minimum cement content in kilograms per cubic metre, the rate of sampling and the number of cubic metres per sample. Optional additional information which is not required by the CESMM includes details of workability, maximum free water/cement ratio, maximum cement content, details of special cements and aggregates, admixtures and testing.

Item descriptions for special prescribed mixes (F 3 * *) must include the permitted type of aggregate and the mix proportions in terms of the weights of cement, coarse aggregate and fine aggregate and the minimum cement content. The rate of sampling and the number of cubic metres per sample are also to be stated. As with designed mixes, various optional statements can be made to elaborate the specification. Only the essential items are required to be stated to comply with rule A1 of class F.

BS 5328 itself encourages the use of a form for scheduling specified requirements of different mixes to be used on one contract. Where each different mix is given a different identifying number or code in accordance with this procedure, this code may be used in the item description in the Bill of Quantities as a mix reference in accordance with rule A1.

The constituents of a 'special prescribed' mix can be almost anything. Note that this is the only type of mix the grade of which is not required by the CESMM to be stated. The CESMM implies that the normal preferred large aggregate sizes of 10 mm, 14 mm, 20 mm and 40 mm apply to this type of mix. If another aggregate size were specified the mix would be classified as F 309 and the particular nature of the aggregate would need to be stated. The grade classification is the same as the 'characteristic compressive strength' defined in BS 5328. This is the 28 day crushing strength expressed in newtons per square millimetre. Flexural or indirect tensile strengths at 28 days may be specified for a special prescribed mix in place of compressive strength. In this, as in other classes of the CESMM, where a concrete grade is preceded by C, the compressive strength is to be used as the grade indicator.

The boundary of coverage of the items for the provision and placing of in situ concrete is not defined. The tenderer may allocate cost between Method-Related Charges and the provision and placing items at his discretion. Estimators will normally allow for the material and mixing cost of concrete against the items for provision, and for the cost of transporting and placing concrete, vibration of concrete and curing against the items for placing. They should bear in mind that the distinctive cost of any additives or special mix characteristics adopted to achieve a certain workability of concrete should therefore be allowed for against the items for the placing of concrete, although the cost incurred is more directly associated with the operation of mixing the concrete. This apparent contradiction will produce a realistic pricing of the two items and therefore lead to the greatest correlation between cost and valuation in any remeasurement of the work.

The measurement rule M1 provides the basic convention for calculating quantities for both providing and placing concrete. In CESMM2, the cross-sectional size of rebates etc. whose volume is ignored when calculating concrete volumes is doubled.

The note printed at the bottom of page 39 of CESMM2 emphasizes the high significance of location to concrete placing costs. It is consequently particularly important that bill compilers should apply paragraph 5.10 thoroughly to items for placing concrete, and give item descriptions and quantities which relate closely to work done in different parts of a structure. Height above ground and plan position are not the only criteria of location which are cost-significant. Often density of reinforcement, difficulty of access, special limitations on rate of pouring and particularly pour size have a larger impact on placing cost. It is good practice to aim to provide additional description or headings identifying the general or particular location of pours for all concrete placing items.

It should be noted that compliance with this note is not mandatory. This is for the technical reason explained in the discussion of paragraphs 2.5, 5.8 and 5.10. Bill compilers should treat it as if it were mandatory because it is worded in weaker terms only so that the bill cannot be regarded to be in error if the rule is applied with less than perfect foresight about likely differences in concrete placing cost. If there are any special requirements for curing placed concrete these should be mentioned to comply with paragraph 5.10.

The classification table only requires the dimensions of concrete components to be given in ranges, not as precise thicknesses or cross-sectional areas. Thus the volumes of all slabs in the same location whose thicknesses are in the range 300–500 mm would be added to give the quantity against a single item. This is a concise example of a characteristic change introduced by the CESMM. Instead of producing separate items for each thickness, even if the difference were only 10 mm, the CESMM produces separate items for work which has different costs due to different locations, but groups ranged thicknesses together. The resulting bill is tuned to the particular job; it shows up significant cost differences and ignores insignificant ones.

All items for the provision and placing of concrete, irrespective of thickness, are measured in cubic metres.

Compilers of bills should apply paragraph 5.14 when dealing with sizes of concrete members, and give the precise thickness or cross-sectional area in place of the range when only one thickness or cross-sectional area is included in one item.

The rules for classification of items for placing concrete do not refer to the thickness of tapering walls or the area of tapering columns and beams. These should be treated as non-standard shapes and numbered E4–6 * 9. Special description should be used to describe the range of specific thicknesses or cross-sectional areas of each tapered component. As this is, in effect, giving more information than the CESMM requires, it is not necessary to treat it as a departure from the CESMM.

Figure 16 is a nomogram to help bill compilers place rectangular concrete components in the right dimension range without calculation.

In situ concrete provides examples of the use of paragraph 5.13 in place of arbitrary boundaries in measurement conventions. The rules in class F require that beams should be separated from columns in bill items, and that slabs should be separated from walls. Beyond a large angle of inclination a component which most people would recognize as a beam becomes more like a column. Similarly, although perhaps at a different angle of inclination, a component which most people would recognize as a wall may be tilted so much that it becomes more like a slab. In accordance with paragraph 5.13, where a steeply inclined component has to be measured and there is uncertainty as to whether it is a beam or a column, for clarity either additional description should be given locating the item with reference to the Drawings or the component should be measured as an 'other concrete form' using rule A4.

This procedure for dealing with components of work which do not fall clearly into one or other of the categories listed in the classification is exemplified in the example bill at item F 5 5 5.1. A column cap might be thought of as a thickening of the slab which it is carrying or as a thickening of the column which is carrying the cap. Under the CESMM rules a column cap could be classed as a thickening to a slab, a thickening to a column or an 'other concrete form'. Paragraph 5.13 is included in the CESMM specifically to deal with this problem. Its effect is to require the bill compiler, when he is in any doubt as to the right category into which to place a particular component, to make what may well be an arbitrary decision and then to give additional description which makes clear what he has done. In the example bill the compiler has arbitrarily decided that the concrete cap is part of the column and has added description to identify the piece of concrete which is covered by the item.

The use of the category 'other concrete forms' provides a conve-

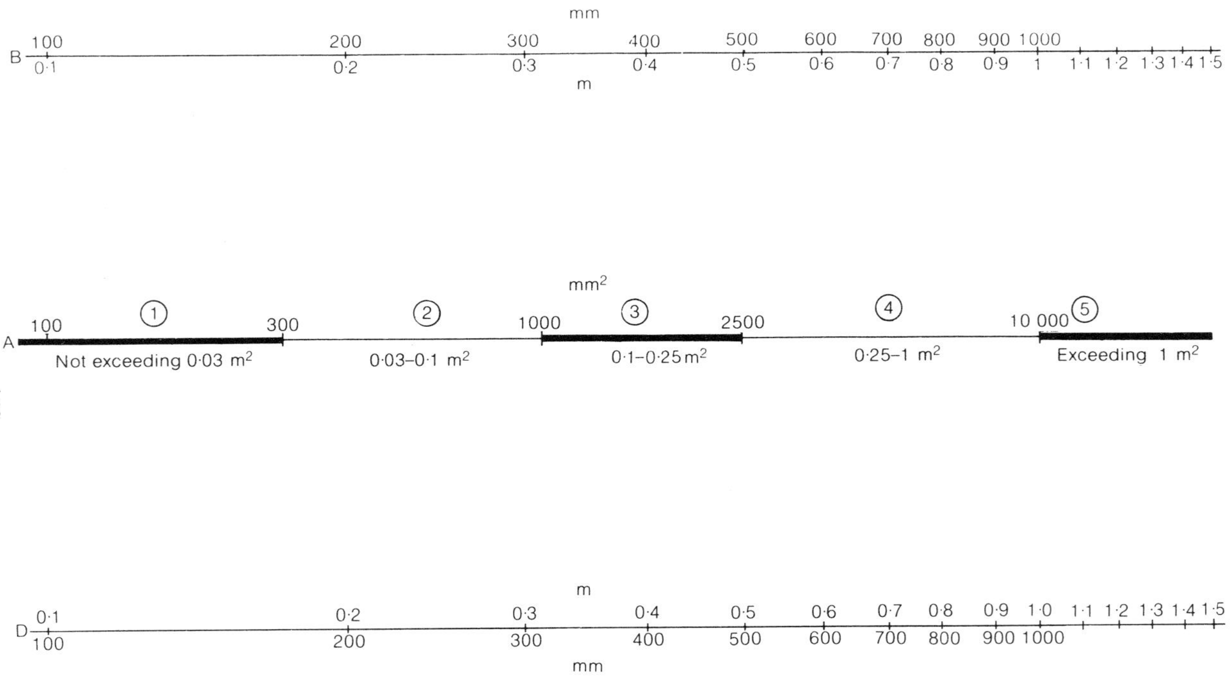

mm
B 100 200 300 400 500 600 700 800 900 1000
0·1 0·2 0·3 0·4 0·5 0·6 0·7 0·8 0·9 1 1·1 1·2 1·3 1·4 1·5
m

mm²
A 100 ① 300 ② 1000 ③ 2500 ④ 10 000 ⑤
Not exceeding 0·03 m² | 0·03–0·1 m² | 0·1–0·25 m² | 0·25–1 m² | Exceeding 1 m²

m
0·1 0·2 0·3 0·4 0·5 0·6 0·7 0·8 0·9 1·0 1·1 1·2 1·3 1·4 1·5
D 100 200 300 400 500 600 700 800 900 1000
mm

Fig. 16. Nomogram for areas of cross-section of rectangular columns, beams, piers and casing to metal sections. If a straight edge is placed from the breadth dimension on scale B to the depth on scale D, the intersection on scale A shows the range from the CESMM in which the cross-sectional area of the component occurs. If the intersection is very close to a range boundary, it is necessary to check by calculation

nient way of describing a composite member which may be composed of walls and slabs or beams and columns. In the example bill a box culvert is described as an other concrete form instead of as a combination of walls and slabs. As use of this category results in fewer items than would be needed using description divided into walls and slabs, it should be used only when the resulting description (which may include a location) is more helpful to the estimator. This alternative should never be used merely as a way of abbreviating a bill or of avoiding detailed measurement of walls and slabs.

The main use of this category will be in dealing with those many volumes of concrete in civil engineering work which are not structural elements of the ordinary building structure type. Theoretically two items could be given for placing all the concrete in a buttress dam—spillways, stilling basins and everything joined on included. An item described as 'Placing mass concrete in Cardingmill Dam' and another 'Placing reinforced concrete in Cardingmill Dam' would satisfy the requirements of rule A4(*c*). Clearly it would be against the spirit and intention of the CESMM, as conveyed by paragraphs 2.5, 5.8, 5.10 and others, to do this. That it is possible demonstrates how much judgement based on knowledge of engineering construction technique is important to the preparation of good bills of quantities using the CESMM.

The category of components called 'Special beam sections' is another example of the same point. The filling of complex beam shapes where a back sloping or top form is required is difficult and relatively expensive. The CESMM therefore requires beams the cross-sectional profiles of which are not approximately rectangular for more than four fifths of their length to be shown separately in the bill as special beam sections with either details of their cross-sectional dimensions or a drawing reference given. This is the effect of rules D8 and A3.

Rules M3, M4 and D6 combine to mean that placing concrete in columns joined on to walls or in beams joined on to slabs will be measured at the rates inserted for the walls or slabs, not at the rates for isolated columns and beams. This is a fail safe arrangement as far as the Contractor is concerned because the placing cost of the concrete in the integral column or beam is likely to be less rather than more than that of the concrete in the adjoining wall or slab. So that the procedure is not 'fail dangerous' for the Employer, the bill compiler should make sure of giving explicit locational descriptions in those cases where a large proportion of the volume of concrete in the resulting item is to be placed in the projections and not in the wall or slab.

Rules M1 and M2 are measurement conventions designed to simplify calculation of quantities. They lead to measurement of a larger volume of concrete than is actually required: some quite large intrusions are not deducted from the concrete volume measured whereas only small projections are not added.

The terms 'internal splays' and 'external splays' are used in rules M1 and M2. An internal splay is one formed by a fillet placed inside formwork reducing the volume of concrete required. An external splay is one which increases the volume of concrete which would have been required had the splay not been cast. The maximum cross-section area of a nib or splay not allowed for in calculating concrete volume is 0·01 $m^2$. This is the area of a square of side length 100 mm or of a 45° splay of side length 140 mm. The maximum volume of cast-in components not deducted is 0·1 $m^3$. This is the volume of a cube of side length 464 mm. The maximum volume set is quite large, not so that the Employer is done out of much concrete, but so that the majority of such components should be clearly less than the limiting volume. It does not simplify calculations if most of the volumes have to be calculated carefully in order to decide whether or not to ignore them.

*Schedule of changes in CESMM2*

1. Specification of concrete mixes is aligned with BS 5328 instead of CP 110.
2. Pile caps are added to the classification of concrete placing.
3. Concrete placed against Excavated Surfaces (other than blinding) is to be identified.

STREFFORD BRIDGE

| Number | Item description | Unit | Quantity | Rate | Amount £ | p |
|---|---|---|---|---|---|---|
| | IN SITU CONCRETE. | | | | | |
| | Provision of concrete. | | | | | |
| F163 | Ordinary prescribed mix grade C30 cement to BS 12, 20 mm aggregate to BS 882. | m3 | 840 | | | |
| | Placing of concrete. | | | | | |
| | Reinforced. | | | | | |
| F534 | Suspended slab thickness: exceeding 500 mm; voided bridge deck. | m3 | 462 | | | |
| F554 | Columns and piers cross-sectional area: 0.25 - 1 m2. | m3 | 96 | | | |
| F564 | Beams cross-sectional area: 0.25 - 1 m2. | m3 | 38 | | | |
| F565 | Beams cross-sectional area: exceeding 1 m2. | m3 | 53 | | | |
| F566.1 | Beams special sections: type J1 drawing 137/26. | m3 | 87 | | | |
| F566.2 | Beams special sections: type J2 drawing 137/26. | m3 | 104 | | | |
| | | | | PAGE TOTAL | | |

INLET WORKS

| Number | Item description | Unit | Quantity | Rate | Amount £ | Amount p |
|---|---|---|---|---|---|---|
| | IN SITU CONCRETE. | | | | | |
| | Provision of concrete, ordinary prescribed mix. | | | | | |
| F113 | Grade C10 cement to BS 12, 20 mm aggregate to BS 882 | m3 | 350 | | | |
| F134 | Grade C15 cement to BS12, 40 mm aggregate to BS 882. | m3 | 1032 | | | |
| F138 | Grade 15 sulphate resisting cement to BS 4027, 40 mm aggregate to BS 882. | m3 | 504 | | | |
| F253 | Provision of concrete designed mix for ornamental surfaces grade C25 cement to BS 12, 20 mm aggregate; minimum cement content 250 kg/m$^3$ sampling as Specification clause 252. | m3 | 632 | | | |
| | Placing of concrete. | | | | | |
| F411 | Mass blinding thickness not exceeding 150 mm. | m3 | 97 | | | |
| F480.1 | Mass backfilling around structure; beneath inlet channel placed against an excavated surface. | m3 | 210 | | | |
| F480.2 | Mass benching; wet wells. | m3 | 21 | | | |
| F480.3 | Mass filling to flume; inlet channel to pumping station. | m3 | 5 | | | |
| F480.4 | Mass plinths; 900 x 900 x 1000 mm. | m3 | 7 | | | |
| F480.5 | Mass plinths; 1500 x 1500 x 1500 mm. | m3 | 10 | | | |
| | | | PAGE TOTAL | | | |

INLET WORKS

| Number | Item description | Unit | Quantity | Rate | Amount £ | p |
|---|---|---|---|---|---|---|
| | IN SITU CONCRETE. | | | | | |
| | Placing of concrete. | | | | | |
| | Reinforced. | | | | | |
| F523.1 | Bases and ground slabs thickness: 300 - 500 mm. | m3 | 906 | | | |
| F523.2 | Ground slabs thickness: 300 - 500 mm; to be placed in one continuous pour in filter block base. | m3 | 15 | | | |
| F533 | Suspended slabs thickness: 300 - 500 mm. | m3 | 120 | | | |
| F541.1 | Walls thickness: not exceeding 150 mm. | m3 | 31 | | | |
| F541.2 | Walls thickness: not exceeding 150 mm; weir wall top finished to precise line and level. | m3 | 10 | | | |
| F542 | Walls thickness: 150 - 300 mm. | m3 | 290 | | | |
| F543 | Walls thickness: 300 - 500 mm. | m3 | 520 | | | |
| F553 | Columns and piers cross-sectional area: 0.1 - 0.25 m2. | m3 | 97 | | | |
| F554 | Columns and piers cross-sectional area: 0.25 - 1 m2. | m3 | 94 | | | |
| F555 | Columns and piers cross-sectional area: exceeding 1 m2; column cap. | m3 | 10 | | | |
| F564 | Beams cross-sectional area: 0.25 - 1 m2. | m3 | 70 | | | |
| F566.1 | Beams special sections; 600 x 900 mm (internal) box beam, wall thickness: 150 mm. | m3 | 67 | | | |
| F566.2 | Beams special sections; type J3 drawing 137/27. | m3 | 76 | | | |
| F566.3 | Beams special sections; type J4 drawing 137/27. | m3 | 98 | | | |
| | | | PAGE TOTAL | | | |

INLET WORKS

| Number | Item description | Unit | Quantity | Rate | Amount £ | p |
|---|---|---|---|---|---|---|
| | IN SITU CONCRETE. | | | | | |
| | Placing of concrete. | | | | | |
| | Reinforced. | | | | | |
| F566.4 | Beams special sections; type J5 drawing 137/27. | m3 | 66 | | | |
| | Other concrete forms. | | | | | |
| F580.1 | Diffuser drum thickness 150 mm, diffuser type D1 drawing 137/49. | m3 | 3 | | | |
| F580.2 | Box culvert internal dimensions 1.5 x 3 m wall thickness: 250 mm. | m3 | 78 | | | |
| F580.3 | Box culvert internal dimensions 2 x 4 m wall thickness: 250 mm. | m3 | 104 | | | |
| F580.4 | Steps and stairs thickness: 100 - 300 mm. | m3 | 12 | | | |
| | | | | PAGE TOTAL | | |

# *Class G: Concrete ancillaries*

### *Formwork*

The rules for measurement of formwork for in situ concrete are given in the classification table at G 1 * * to G 4 * *.

Rule M1 is the main rule for measuring formwork. It is qualified by other rules. The rules sometimes result in instances where formwork is measured although not provided and other instances where formwork is not measured although provided. Formwork is a distinctive problem in this respect as it is without question Temporary Works although it is traditionally given in Bills of Quantities as if it were Permanent Works.

Rule M1 applies 'except where otherwise stated in the CESMM'. Class G specifically excludes formwork for work included in classes H, K, L, P, Q, R, T and X. Formwork generally is not measured for in situ concrete which is included in these classes. Rule M1 consequently does not apply to formwork for work excluded from class G; it does not mean, for example, that formwork to thrust blocks in pipework or to beds and backings to kerbs in road-works is measured. The exceptions to the general rule M1 are given in the measurement rule M2.

Rule M2(*a*) states that formwork to blinding concrete not exceeding 0·2 m wide is not measured. Its cost is to be included in other rates entered in the Bill of Quantities. In CESMM2 formwork to blinding concrete which exceeds 0·2 m wide or deep is measured in accordance with the rules for formwork generally and is separately described in accordance with rule A3. The phrase 'wide or deep' is used here, as it is in the description of joints, to mean the narrower dimension of a surface.

Rule M2(*b*) has the effect that items for joints measured in accordance with the classification table at G 6 * * include the costs associated with forming the joint.

Rule M2(*c*) states that no formwork items are measured to temporary surfaces formed at the discretion of the Contractor. These may include the surfaces of joints which exist only between successive concrete pours.

Rule M2(*d*) states that, where concrete is expressly required to be cast against an excavated surface, no formwork is to be measured. If such a requirement is not expressed, formwork is deemed to be required. Where the concrete is expressly required to be cast against an excavated surface, rule M11 of class E requires preparation of the excavated surface to be measured.

Rule M2(*e*) establishes that formwork is not measured to surfaces inclined at an angle exceeding 45° to the vertical which are cast against excavated surfaces (even if not expressly required to be so cast). This rule is provided so that a distinction can be drawn between the bottom of something concrete for which formwork should not be measured and the side of it for which formwork should be measured. This distinction is illustrated in Fig. 17. The surface AB is regarded as a side; the surface BC as a bottom.

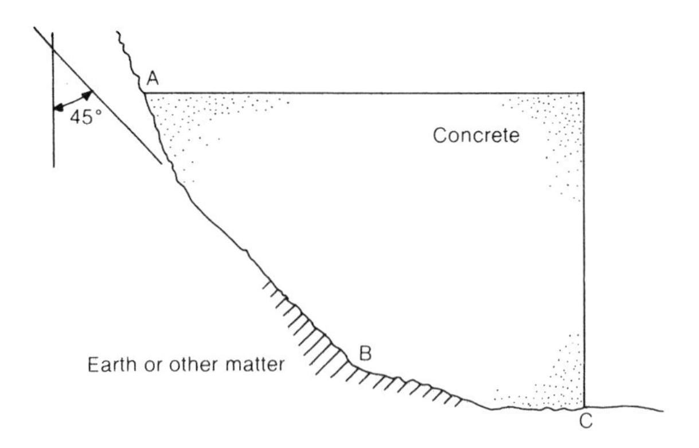


Fig. 17. Illustration of rule M2(*e*). Formwork is measured for the surface AB but not for the surface BC

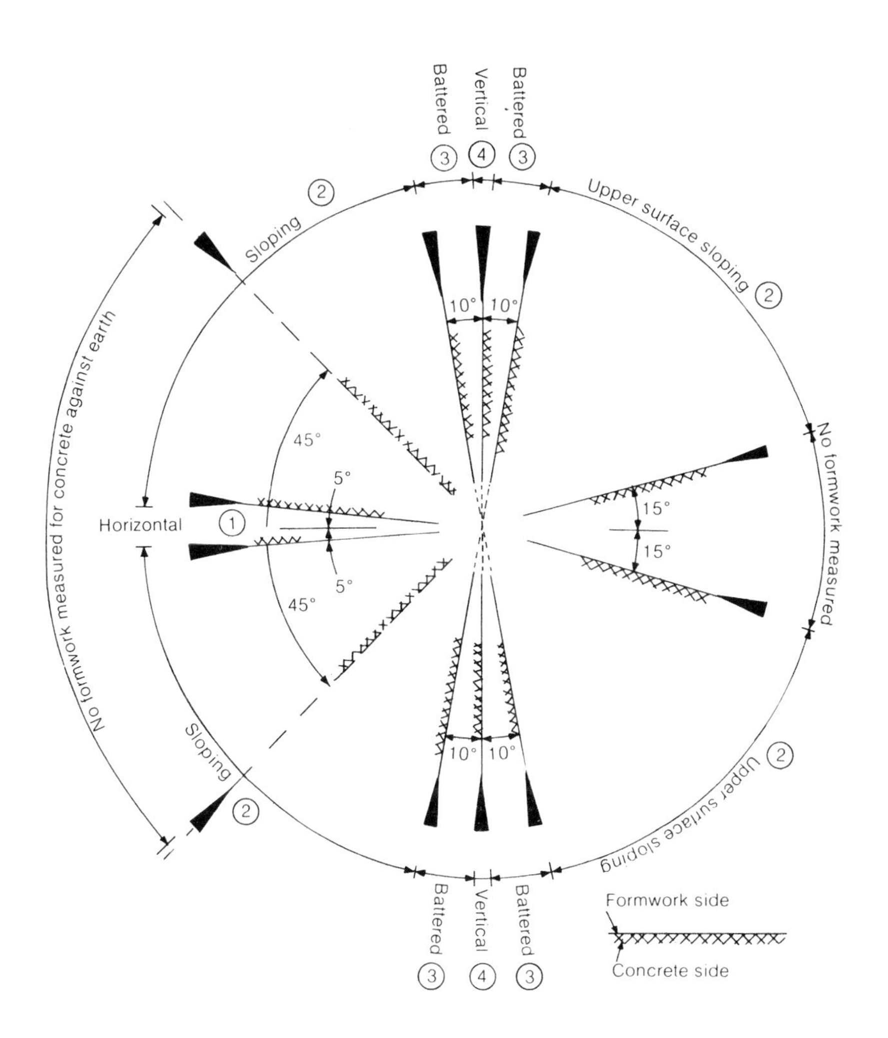


Fig. 18. Inclination zones for plane formwork defined in rules M2(*e*), M3, D1 and A2. Note the precise boundaries of the zones. For example, an inclination of 10° to the vertical is in zone 3; an inclination of $10\frac{1}{2}$° is in zone 2

Rule M3 establishes that upper surfaces of concrete are measured for formwork only where they are inclined at an angle exceeding 15° to the horizontal or where it is expressly required that other upper surfaces shall be formed. In the absence of an express requirement, the cost of optional upper surface formwork such as at thresholds of wall openings must be allowed for in other items. Strictly this means that the 30° bottom segment of a circular void former mounted horizontally would not be measured. To avoid this sort of pedantry, work of that type should be measured linearly as 'components of constant cross-section' (items G 1–4 8 9) although the void former does not necessarily produce a concrete member of constant cross-section. Upper surfaces are required to be described separately so that the different arrangements for support which they require can be allowed for in pricing (rule A2).

Rule D1 gives the angles of inclination which bound the categories of horizontal, sloping, battered and vertical plane formwork. Note that it applies only to plane and not to curved formwork. The effects of rules M2(*e*), M3, D1 and A2 are illustrated in Fig. 18. This shows the appropriate category of formwork for a plane concrete surface inclined at any angle. The additional description 'upper surface' formwork required by rule A2 is applied only to surfaces classed as sloping. A copy of Fig. 18 on clear paper can be used by the bill compiler to categorize concrete surfaces directly from the drawings.

The note at the foot of page 41 of CESMM2 explains where formwork may be measured by length as one item, instead of by area in items for the separate surfaces to be formed. This is where concrete members (or holes in concrete members) are of constant cross-section. In such cases (for example beam and column boxes and travelling forms for walls) the formwork will normally be made up so that it can be re-used several times without the necessity for breaking down and re-assembling the separate surfaces. Where this alternative is used, the bill compiler must give additional description as required by rule A5 which identifies the concrete members by their principal dimensions and a mark number or other reference. Items G 2 8 1–4 in the example bill demonstrate this procedure. It should be used freely and not as an inferior option. It draws attention to situations where many re-uses of formwork may be possible and produces a price in the Contract which reflects the realities of the cost of formwork. It is almost essential to use it for complex cross-sectional shapes such as I beams, box beams and fluted columns. If a large I beam were not to be measured using this procedure, the area of formwork for its surfaces would be divided between six items, adding an amount to the total quantity for the formwork which was collected in each item. The particular significance and cost characteristics of the I beam formwork would be obscured completely. The constant cross-section procedure is not confined to beams and columns. It can be used for walls, parapets and for circular or curved members provided that their cross-section is constant as it affects the formwork. Any member embodying intermittent or irregular projections cannot be classed as a component of constant cross-section. Strictly speaking neither can one with intermittent or irregular pockets or voids, but as these do not affect re-use of the main formwork they should still be measured in this way.

Rule A1 is a new rule in CESMM2 which requires permanent formwork to be identified and given in separate items. It does not distinguish between formwork left in because it is expressly required to be left in and formwork which it is impossible to remove.

The first classification of formwork given in class G is by the different surfaces required. It is sometimes helpful to define rough and fair finishes in terms of the anticipated qualities of surface in a Preamble clause in the same way as rock is defined in terms of the anticipated geological formations in the particular job. The four types of formwork are alternatives; surface features are not measured in addition to any ordinary backing formwork to which they are secured.

The distinction between a 'finish' and a 'surface feature' is not stated. Obviously a feature is something in higher relief than a finish and anything which has to be fixed to formwork to create a shaped decorative feature is more likely to be called a feature than a finish. Fortunately the distinction has no contractual importance, as the particular surface detail has to be stated in the item description in either case. Closely spaced projections and intrusions can be classed as 'surface features' when appropriate.

The classification by width of formwork in the third division is intended to differentiate between areas of formwork involving different ratios of labour cost to the measured area. All narrow widths and other longitudinal features of formed surfaces which do not exceed 200 mm wide other than projections and intrusions are measured linearly in items G 1–4 1–5 1–2. Although no distinction is made between the various complex surfaces which may be compounded from such narrow widths, the total length of the various features will indicate the general complexity of small features associated with particular formwork assemblies. Where kickers for walls are expressly required, the formwork will be measured as the appropriate width and included with the other narrow surfaces. The width used to distinguish between items G 1–4 1–5 4 and G 1–4 1–5 5 is the normal full panel plywood width which was formerly a nominal 4 ft in imperial measure (now 1·22 m). The distinction is therefore between formwork surfaces which are less than one panel wide (items G 1–4 1–5 4) and those which may comprise one or more full panel widths with a make-up piece (items G 1–4 1–5 5).

CESMM2 has a new way of dealing with narrow widths of formwork comprising ribs and external splays (defined in rule D4 as projections) and rebates, grooves, internal splays, throats, fillets and chamfers (defined in rule D5 as intrusions). Where any of these have a cross-sectional area not exceeding 0·01 $m^2$, they are lumped together and given as one item for all projections (G 1–4 8 5) and one for all intrusions (G 1–4 8 6). Projections and intrusions dealt with in this way for formwork are now the same surface features as those which are ignored for concrete volume calculations in class F (rules M1(*d*) and M2 of class F).

Rule D2 has the effect of abbreviating item descriptions by eliminating the need to repeat the words 'plane' and 'width: exceeding 1·22 m' in item descriptions; they will be assumed unless contradicted. The example bill items demonstrate this, particularly in items G 1 4 5, G 2 1 5, G 2 2 5 and G 2 3 5.

Rule A4 provides the rule for dealing with curved formwork. It can be summarized as requiring separate items to be given for each different radius of formwork and each different shape of multi-radius formwork. Curved formwork is not classified by inclination. It is clearly helpful if location is indicated in item descriptions for curved formwork. As cutting of formwork is not measured it is not necessary to draw attention to curved cutting, i.e. cutting of formwork which is a flat surface but which has a curved edge. Making complex or expensive formwork, particularly curved formwork, will often be priced by tenderers as a Fixed Charge in class A. The classification table requires items for single radius (cylindrical) formwork to be classified according to width. Items for other curved formwork are not classified according to width. In CESMM2 a new rule A6 requires components of constant cross-section which are themselves curved to be described with their radii stated.

Rules M4, M6 and D3 provide the rules for measuring formwork to voids. Voids are not defined in the CESMM. The natural meaning of 'formwork to voids' is formwork placed in order to form a hole into, within or through an in situ concrete member. This meaning fits the use of the term made by the CESMM. Any self-contained pocket or hole formed by attaching a box or former to the inside of a form should be classed as a void. Pockets should be measured where they are required to be formed, but not where the Contractor leaves a temporary pocket at his own discretion. This occurs where inserts are to be cast in but the Contractor is allowed to box out for them and to grout the inserts into the pockets at a later stage.

Voids are not measured in detail; the number required is given in two ranges of cross-section size and three ranges of depth not exceeding 2 m, the depth being stated for voids deeper than 2 m. The formwork for any void whose cross-sectional area exceeds 0·5 m$^2$ or whose diameter exceeds 0·7 m is treated as ordinary formwork. The top of the size range called 'small' includes square cross-section voids with sides of length up to 316 mm and circular cross-section voids with diameter up to 350 mm. The depth of voids used for classification is the perpendicular distance from the surface where the void is exposed to the end of the void. The end of the void may be another surface if the void goes right through a concrete member. In the unusual case where the surfaces at each end of a void are not parallel the rule for measuring depth (rule D3) produces two possible lengths. Bill compilers should be alert to this and should indicate the surface from which the depth has been taken if it would affect the depth range.

Note that the volume of concrete displaced by a large or small void is not deducted from the volume of concrete calculated (rule M1(*e*) in class F). This rule applies irrespective of the depth of the void. Tenderers should make sure that their price reflects this. A large void can displace up to half a cubic metre of concrete per metre of depth. The rates for large voids should take this into account if the tender is to be priced realistically.

The approach to measurement of formwork adopted in the CESMM relies on the tenderer deriving much of the information necessary to estimate the cost of formwork from drawings of the concrete structures to be cast. No bill item description could convey all the information which is needed to determine the support system, the pour heights, the striking procedure and the number of re-uses of formwork. Bill compilers will facilitate good planning of formwork systems and accurate estimating by giving full locational information in formwork item descriptions. This is another example of the advantages to be gained by thorough application of paragraph 5.10. Any variations to formwork are dealt with more easily and equitably if the estimated fixed and time-related costs are allowed for in the tender by Method-Related Charges.

Tolerances are often a vital influence on the cost of erecting and fixing formwork. Consequently bill compilers should be particularly careful to draw attention to different or special tolerance requirements in accordance with paragraph 5.10.

### *Reinforcement*

The rules for measurement of reinforcement are given in the classification at G 5 * *. The measurement rules are quite traditional. Supply and fixing of reinforcing steel is still a relatively unmechanized operation needing no expensive Temporary Works or specialized site plant. It is consequently well suited to the traditional quantity proportional measurement of the amount of material left behind in the Permanent Works when the work is finished.

The rules generate separate items for work in structures in different bill parts and work in different reinforcing materials of different sizes. Reinforcing bars are classified by nominal size, otherwise diameter taken from the preferred sizes for steel bars referred to in BS 4449.* In CESMM2 the size listing has been changed so that sizes of less than 32 mm are each given separately and sizes of

* *Hot rolled steel bars for the reinforcement of concrete*. British Standards Institution, London, 1969, BS 4449.

32 mm and greater are grouped in one item. All other lists of bar reinforcement sizes in CESMM2 have been altered similarly.

Rule C1 refers to supporting reinforcement. This means the work and materials needed to support, tie and otherwise to secure the reinforcement. Only the cost of additional material introduced to support top reinforcement is allowed for by including the mass of this material in the mass of reinforcing steel itself. This rule is not confined to the amount of such steel expressly required. It is consequently measured in accordance with rule M8 for the amount of support steel the Contractor provides, even if none is shown on the Drawings. The Contractor has no incentive to overdo the support unless the rates for ordinary reinforcement are especially lucrative—which they will not be unless the Contractor correctly deduced that the reinforcing steel in the original bill was significantly under-measured at the time of tender.

Rule A7 is included so that both the supplier's 'extra' for long bars and the additional cost of handling and fixing long bars can be reflected in the prices. Where no length is stated it will be assumed to be 12 m or less before bending. Where it is stated as 15 m, 18 m or 21 m this will be assumed to mean 12–15 m, 15–18 m and 18–21 m respectively. This is shown in the descriptions of items G 5 1 5.1, G 5 1 6.1 and G 5 1 6.2 in the example bill.

CESMM2 introduces new items for special joints in bar reinforcement as defined in rule D7. The items (at G 5 5 0 in the classification table) are to be described in accordance with rule A8.

Rule A9 does not require details of fabric reinforcement complying with BS 4483* to be given in item descriptions. This is because the type number taken from BS 4483 is equivalent to these details and it is required to be stated instead. In accordance with rule M9 the area of additional material in laps in fabric reinforcement is not measured. This is the usual practice which is well understood.

## *Joints*

Measurement of movement joints is divided into two types of item: those for the work associated with the area of the joint (G 6 1–

* *Steel fabric for the reinforcement of concrete.* British Standards Institution, London, 1969, BS 4483.

4 *) and those associated with the length of some component of the joint (G 6 5–7 *). The area items are taken to cover formwork (rule C3) and the cost of supplying and fitting filler material. The length items are of two types covering internal detail and face detail. In CESMM2 these two types of detail are not referred to separately but items are given for them without change from the first edition.

The measurement of formwork to joints was a source of confusion to some users of the first edition of the CESMM. The confusion centred on the meaning of note G14 which was said to be ambiguous. To remove any hint of ambiguity, CESMM2 now contains two statements on the point. Rule M2(*b*) of class G says that formwork shall not be measured for joints and their associated rebates and grooves. Rule C3 of class G says that items for joints are deemed to include formwork.

The rule for when to measure joints in concrete is also made more explicit in CESMM2. Rule M10 says that they are only measured where they are at locations where they are expressly required.

It is important when preparing tender documents not to build in ambiguity on this point. For example, joints shown on drawings will be assumed to be expressly required unless specifically annotated otherwise. Similarly, guidance on joint location or constraints on joint location given in the Specification or on drawings can leave the tenderer unsure of whether the joints to be put in at the locations which are implied will be regarded as having been expressly required or not. The words used in specifications or on drawings should be checked on this point and modified, if necessary, to avoid ambiguity before being issued for tendering.

The temporary surfaces referred to in rule M2(*c*) are contractors' day joints or construction joints. They are temporary in the sense that they have no permanent function and, except in water-retaining work, in that there is no material dividing one side of the joint from the other in the completed work.

The distinction between open surface and formed joints is defined in rule D8. It means generally that horizontal surface joints are open surface joints because they do not need stop-end formwork and that vertical surface joints are formed joints because they do. The formal definition in the CESMM cannot be worded so simply because it has also to cover the situation where a joint surface is inclined.

As joint fillers are measured by area the prices for them realistically cover the material cost which is proportional to area. Further itemization is needed only to denote ranges of width reflecting the different proportions of fixing and cutting labour cost. For this reason the joint surface classification in the third division is only by ranges of width. The classification of waterstops is also ranged in width when it does not exceed 300 mm although this will often be overridden by the application of paragraph 5.14 so that actual widths are stated. The ranges are fairly narrow to produce this effect frequently, as shown in the example bill. A new rule in CESMM2 (rule C4) clarifies the position that items for waterstops cover cutting, joining, angles and junctions.

Rule A11 requires a comprehensive description of composite internal joint details such as dowel assemblies to be given. Spacing of dowel bars and their dimensions and thicknesses of filler materials should also be given. Rule M11 establishes a simplifying measurement convention that all joint surface areas and classifications of width shall be determined from the full width of the concrete member. A joint such as a crack control joint which penetrates only a short distance from one surface should be classed as a surface detail with the distance stated as a dimension in accordance with rule A11. The reference to average width or depth in rule D9 is included to allow for tapering joints.

*Post-tensioned prestressing*

The measurement of post-tensioned prestressing is very simple as it relies on the fact that detailed specifications of stressing components and procedures are normally given and that profiles and positions of stressing tendons can only be indicated effectively on the Drawings. The rules are unaltered in CESMM2. Rule A12 dominates the rules in requiring item descriptions to identify separately the different concrete members to be stressed and to give details of the components required. The classification table for prestressing has little impact on the pricing of the items. Estimators will continue to have to work up prestressing prices from their examination of the Drawings and Specification and their assessment of the quantities of ancillary items such as the ducts and vent pipes which are required.

A general point should be made in this context. Prestressing is not particularly prone to variations as regards redesign of the details of systems. It is conceivable that the lengths of stressing tendons might be changed as a result of a variation, for example from 15·1 m to 20 m. This would produce a 30% increase in the material cost of tendons and ducts although the resulting tendons would still be within the same length range given in the CESMM. This should not be taken as meaning that the original rate is necessarily still applicable and reasonable. The item descriptions in the bill identify work and if the work identified changes so may the bill rate. The test of whether or not the work has changed is whether or not it differs from that originally drawn and specified, not whether or not it can still be held to be within the possible interpretations of the original item description. Item descriptions identify work; they do not define it. This general point applies equally to those variations which reduce the cost of work as to those which increase it.

*Concrete accessories*

Finishes applied to in situ concrete by means other than casting against formwork which carries the surface finish or features required are measured according to rules M13, M14, C6 and A13 and the classification table at G 8 1–2 *. The items for top surface finishes include applied finishing layers such as granolithic and other screeds. The thicknesses of applied layers and the limits of any variations in nominal thickness required to create falls are to be stated. As the thicknesses and materials to be used are stated, their mix details and volumes should not be included in the item descriptions and quantities given in class F for normal in situ concrete.

Rule M13 provides that no finishing to top surfaces of walls or to other concrete pours should be measured if no separate finishing is required. A separate treatment is any finishing treatment requiring specialist labour to do something after a pour is finished, not normal levelling off carried out by the concrete placing gang themselves. The items for finishing formed surfaces (G 8 2 *) should not be used to measure remedial work to surface finishes left imperfect after stripping formwork, even if the prospect of remedial work is referred to in the Specification.

Class G includes a classification and rules of measurement for inserts cast or grouted into in situ concrete (G 8 3 *). The arrangements for inserts in the first edition have been substantially expanded and made more precise in CESMM2. The classification table is unaltered but the rules are revised.

Rules M15 and A16 establish the circumstances in which formwork is measured for boxing out before grouting in inserts. Rule D12 defines that everything except reinforcement, prestressing and joints

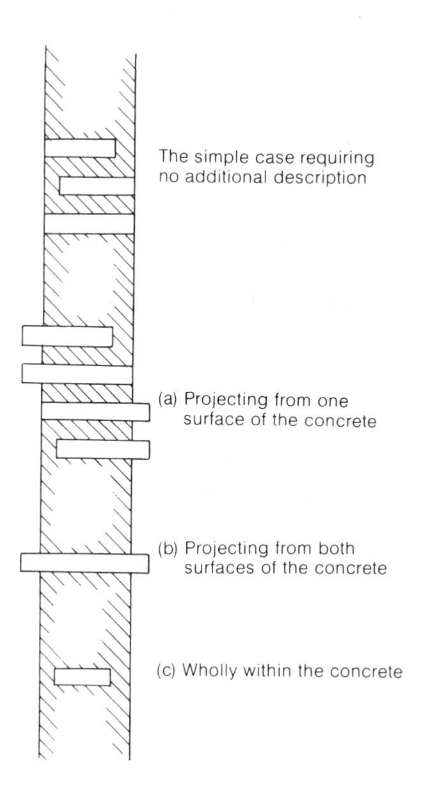


Fig. 19. Inserts classified according to rule A15. The four types of insert are shown illustrating their different effects on wall formwork. The figure viewed on its side illustrates the same points for slab formwork

which ends up cast or grouted into the concrete is an insert. The classification divides these into long inserts (such as anchor slots and steel framing) which are measured linearly and the others (such as anchor bolts and pipe sleeves) which are enumerated. The classification assumes that normally the Contractor will be given the choice of how to fix the inserts into the concrete, whether by casting in, by subsequent grouting into a cast pocket or by drilling and either grouting or using an adhesive.

Rule C7 says that items for inserts include supply unless otherwise stated. In practice, 'otherwise' will often be stated as the supply of many inserts will be covered by items in other classes. The new statement in paragraph 3.3 of CESMM2 addresses this important point.

Rule A15 requires three special cases of inserts to be identified when they occur because of the different effect they have on the cost of fixing inserts to formwork and adapting formwork. The special cases and the general case are illustrated on Fig. 19.

CESMM2 adds items at G 8 4 1–4 for grouting under base plates.

## *Schedule of changes in CESMM2*

*Formwork*

1. Formwork to blinding concrete is measured (other than edges less than 0·2 m deep).
2. Permanent formwork is to be identified.
3. Formwork for projections and intrusions is measured more simply.
4. Joint formwork is not measured but is to be included in joint items.

*Reinforcement*

5. Size 25 mm is given separately.

*Joints*

6. Items include formwork.

*Inserts*

7. Described in more detail.
8. Grouting under base plates is added.

| Number | Item description | Unit | Quantity | Rate | Amount £ | p |
|---|---|---|---|---|---|---|
| | CONCRETE ANCILLARIES. | | | | | |
| G112 | Formwork rough finish horizontal width 0.1 - 0.2 m. | m | 129 | | | |
| G113 | Formwork rough finish horizontal width 0.2 - 0.4 m to be left in. | m2 | 7 | | | |
| G145 | Formwork rough finish vertical. | m2 | 370 | | | |
| G155 | Formwork rough finish curved to one radius in one plane; radius 10.5 m, back of wall D drawing 137/6. | m2 | 997 | | | |
| G184 | Formwork rough finish for concrete component of constant cross-section; box culvert internal dimensions 1.5 x 3 m wall thickness: 250 mm. | m | 326 | | | |
| | Formwork fair finish. | | | | | |
| G211 | Horizontal width not exceeding 0.1 m. | m | 410 | | | |
| G213 | Horizontal width 0.2 - 0.4 m. | m2 | 331 | | | |
| G215 | Horizontal. | m2 | 981 | | | |
| G222.1 | Sloping width 0.1 - 0.2 m. | m | 723 | | | |
| G222.2 | Sloping width 0.1 - 0.2 m; upper surface. | m | 27 | | | |
| G225.1 | Sloping. | m2 | 128 | | | |
| G225.2 | Sloping; upper surface. | m2 | 157 | | | |
| G235 | Battered. | m2 | 429 | | | |
| G241 | Vertical width not exceeding 0.1 m. | m | 410 | | | |
| G242 | Vertical width 0.1 -0.2 m. | m | 924 | | | |
| G243 | Vertical width 0.2 - 0.4 m. | m2 | 192 | | | |
| G245 | Vertical. | m2 | 780 | | | |
| | | | | PAGE TOTAL | | |

| Number | Item description | Unit | Quantity | Rate | Amount £ | p |
|---|---|---|---|---|---|---|
| | CONCRETE ANCILLARIES. | | | | | |
| | Formwork fair finish, | | | | | |
| G252 | Curved to one radius in one plane width 0.15 m; radius 1.0 m, column plinths. | m | 927 | | | |
| G255 | Curved to one radius in one plane radius 11.5 m; face of wall D drawing 137/6. | m2 | 1354 | | | |
| G260.1 | Spherical radius 2 m; flume chamber. | m2 | 43 | | | |
| G260.2 | Conical maximum radius 600 mm minimum radius 300 mm; outlet chamber. | m2 | 27 | | | |
| G260.3 | Conical maximum radius 800 mm minimum radius 400 mm; inlet chamber. | m2 | 35 | | | |
| G271 | For small voids depth not exceeding 0.5 m. | nr | 157 | | | |
| G272 | For small voids depth 0.5 - 1 m. | nr | 298 | | | |
| G276 | For large voids depth 0.5 - 1 m. | nr | 76 | | | |
| G278.1 | For large voids depth 2.3 m. | nr | 5 | | | |
| G278.2 | For large voids depth 2.6 m. | nr | 1 | | | |
| | For concrete components of constant cross-section. | | | | | |
| G281.1 | Beams; 200 x 300 mm beam B27. | m | 37 | | | |
| G281.2 | Beams; 300 x 400 mm beam B28. | m | 50 | | | |
| G281.3 | Beams; 450 x 600 mm beam B30. | m | 46 | | | |
| G282.1 | Columns; 450 x 450 mm columns 1 - 24. | m | 98 | | | |
| G282.2 | Columns; 600 x 600 mm columns 25 - 30. | m | 27 | | | |
| | | | PAGE TOTAL | | | |

| Number | Item description | Unit | Quantity | Rate | Amount £ | p |
|---|---|---|---|---|---|---|
| | CONCRETE ANCILLARIES. | | | | | |
| | Formwork fair finish. | | | | | |
| | For concrete components of constant cross-section. | | | | | |
| G282.3 | Columns; curved to 150 mm radius columns 31 - 54. | m | 95 | | | |
| G284 | Box culvert internal dimensions 1.5 x 3 m wall thickness 250 mm beneath entrance road. | m | 310 | | | |
| G285 | Projections. | m | 2076 | | | |
| G286 | Intrusions. | m | 1594 | | | |
| G299 | For diffuser drum including perforations as detailed on drawing 137/54. | m | 5 | | | |
| | | | PAGE TOTAL | | | |

| Number | Item description | Unit | Quantity | Rate | Amount £ | p |
|---|---|---|---|---|---|---|
| | CONCRETE ANCILLARIES. | | | | | |
| | Reinforcement. | | | | | |
| | Mild steel bars to BS 4449. | | | | | |
| G511 | Diameter 6 mm. | t | 3.7 | | | |
| G512 | Diameter 8 mm. | t | 3.9 | | | |
| G513 | Diameter 10 mm. | t | 4.2 | | | |
| G514 | Diameter 12 mm. | t | 4.9 | | | |
| G515.1 | Diameter 16 mm. | t | 5.7 | | | |
| G515.2 | Diameter 16 mm; length 15 m. | t | 2.1 | | | |
| G516.1 | Diameter 20 mm. | t | 5.9 | | | |
| G516.2 | Diameter 20 mm; length 15 m. | t | 1.0 | | | |
| G516.3 | Diameter 20 mm; length 18 m. | t | 0.9 | | | |
| G517.1 | Diameter 25 mm. | t | 6.3 | | | |
| G517.2 | Diameter 25 mm; length 15 m. | t | 1.2 | | | |
| G518 | Diameter 32 mm or greater. | t | 7.6 | | | |
| G526.1 | High yield steel bars to BS 4449 diameter 20 mm. | t | 4.2 | | | |
| G526.2 | High yield steel bars to BS 4449 diameter 20 mm; length 15 m. | t | 2.1 | | | |
| G528 | High yield steel bars to BS 4449 diameter 32 mm or greater. | t | 3.9 | | | |
| G563.1 | High yield steel fabric to BS 4483 nominal mass 3 - 4 kg/m2; type number A193. | m2 | 3728 | | | |
| | | | PAGE TOTAL | | | |

| Number | Item description | Unit | Quantity | Rate | Amount | |
|---|---|---|---|---|---|---|
| | | | | | £ | p |
| | CONCRETE ANCILLARIES. | | | | | |
| | Reinforcement. | | | | | |
| G563.2 | High yield steel fabric to BS 4483 nominal mass 3 - 4 kg/m2; type number A252. | m2 | 1807 | | | |
| G564 | High yield steel fabric to BS 4483 nominal mass 4 - 5 kg/m2; type number C503. | m2 | 286 | | | |
| | Joints. | | | | | |
| | 25 mm bitumen impregnated fibreboard filler. | | | | | |
| G621 | Open surface width: not exceeding 0.5 m. | m2 | 97 | | | |
| G622 | Open surface width: 0.5 - 1m. | m2 | 38 | | | |
| G641 | Formed surface width: not exceeding 0.5 m. | m2 | 107 | | | |
| G642 | Formed surface width: 0.5 - 1 m. | m2 | 98 | | | |
| G643.1 | Formed surface width: 1.2 m. | m2 | 27 | | | |
| G643.2 | Formed surface width: 1.45 m. | m2 | 41 | | | |
| G651 | P.V.C. centre bulb waterstop; 100 mm wide. | m | 38 | | | |
| G652 | P.V.C. centre bulb waterstop; 175 mm wide. | m | 110 | | | |
| G653 | Rubber centre bulb waterstop; 225 mm wide. | m | 150 | | | |
| G654.1 | Rubber centre bulb waterstop; 400 mm wide. | m | 28 | | | |
| G654.2 | Rubber centre bulb waterstop; 450 mm wide. | m | 64 | | | |
| | | | PAGE TOTAL | | | |

| Number | Item description | Unit | Quantity | Rate | Amount £ | p |
|---|---|---|---|---|---|---|
| | CONCRETE ANCILLARIES. | | | | | |
| | Joints. | | | | | |
| G670.1 | Rebate sealed with 25 x 25 mm Pliastic or similar approved. | m | 125 | | | |
| G670.2 | Rebate sealed with 25 x 25 mm Plastijoint or similar approved. | m | 670 | | | |
| G682.1 | Sleeved and capped dowels; 25 mm diameter by 300 mm mild steel bars at 0.5 m centres as detail E drawing 137/25. | nr | 128 | | | |
| G682.2 | Sleeved and capped dowels; 25 mm diameter by 300 mm mild steel bars at 1 m centres as detail E drawing 137/25. | nr | 31 | | | |
| | Concrete accessories. | | | | | |
| G811 | Finishing of top surfaces wood float. | m2 | 4976 | | | |
| G812 | Finishing of top surfaces steel trowel. | m2 | 10821 | | | |
| G814.1 | Finishing of top surfaces granolithic finish; as Specification clause G8/05, 40 mm thick; steel trowel surface treatment. | m2 | 192 | | | |
| G814.2 | Finishing of top surfaces granolithic finish; as Specification clause G8/05, 55 mm thick; steel trowel surface treatment. | m2 | 141 | | | |
| G815 | Finishing of top surfaces cement and sand screed as Specification clause G8/12, 40 mm thick; steel trowel surface treatment. | m2 | 98 | | | |
| G822 | Finishing of formed surfaces bush hammering. | m2 | 487 | | | |
| | | | | PAGE TOTAL | | |

| Number | Item description | Unit | Quantity | Rate | Amount £ | p |
|---|---|---|---|---|---|---|
| | CONCRETE ANCILLARIES. | | | | | |
| | Inserts. | | | | | |
| G831 | 100 mm diameter G.V.C. pipe; excluding supply of the pipe. | m | 27 | | | |
| G832.1 | 10 mm diameter, 25 mm deep expanding bolt projecting from one surface. | nr | 110 | | | |
| G832.2 | 15 mm diameter, 40 mm deep rag bolt projecting from one surface. | nr | 110 | | | |
| G832.3 | 25 mm diameter, 100 mm long mild steel holding down bolt projecting from one concrete surface for scraper machinery as detail L drawing 138/17. | nr | 64 | | | |
| G832.4 | 100 mm diameter ductile iron pipe projecting from one surface; excluding supply of pipe. | nr | 10 | | | |
| G832.5 | 150 mm diameter ductile iron pipe with puddle flange projecting from two surfaces; excluding supply of pipe. | nr | 3 | | | |
| G832.6 | 225 mm diameter ductile iron pipe projecting from two surfaces grouted with grout type G3 into 450 x 450 mm preformed opening; excluding supply and fixing of the pipe under separate contract. | nr | 4 | | | |
| G842 | Grouting under plates area 0.1 - 0.5 m2 with type G2 grout. | nr | 8 | | | |
| | | | PAGE TOTAL | | | |

# *Class H: Precast concrete*

The approach to measurement adopted in class H is somewhat different from that used in other classes. The measurement rules produce bills of quantities which merely count identical precast concrete units of which the details are shown on the Drawings. The rules and the classification table do not require details to be given in the item descriptions of formwork, reinforcement, jointing methods and materials or fixing methods and materials. The tenderer will expect to use the position in the Works and mark number stated in the descriptions to lead him to the places on the Drawings and in the Specification where the detail of what is to be priced is shown.

This approach is founded on the fact that the cost of precast concrete work is highly dependent on manufacturing cost, and that this is a function of the facilities of the manufacturer. As for fabricated structural metalwork, in precast concrete work design which is specific to a job has a big influence on manufacturing cost. This is distinct from pipework, for example, where manufacturing is undertaken for stock independent of the design of particular jobs. The precast concrete components included in class H are those which are designed for particular civil engineering projects. They do not include standard products such as kerbs, manholes, piles and fence posts.

The main characteristic of the larger special precast concrete components in class H is that their cost depends on their shape and size and on how many identical components are required. No description could convey all this as effectively as a drawing, and no rule could guarantee that the particular aspects of shape and size relevant to any one manufacturer's facilities would be stated in the description.

For these reasons the CESMM does not require comprehensive

description of precast units to be given. Instead it requires separate items to be given for each different mark or type number of unit and each differently shaped unit to be given a different mark or type number (rule A2). The example bill items show the descriptions and headings generated by these rules. Notice that the effect of rule A2 is to override the ranged classifications of length, area and mass in the second and third divisions of the classification table. Rule A2 requires the mark number of each unit to be stated; paragraph 3.9 establishes that separate items are consequently required for units with different mark numbers. Rule A2 requires units with different dimensions to have different mark numbers; hence they must also be given in separate items. Paragraph 5.14 says that if an item contains components all having the same dimensions, those dimensions shall be stated in place of the ranged dimensions given in the classification tables.

Taken together, these rules require the following mark numbers and dimensions to be given in item descriptions for precast concrete units

| | |
|---|---|
| *Beams and columns* | Mark number<br>Principal dimensions of cross-section<br>Length<br>Mass |
| *Slabs* | Mark number<br>Average thickness<br>Area<br>Mass |
| *Segmental units and units for subways, culverts and ducts* | Mark number<br>Principal dimensions of cross-section<br>Mass per metre |
| *Copings, cills and weir blocks* | Mark number<br>Principal dimensions of cross-section<br>Cross-sectional area<br>Mass per metre |

Where the lengths of units are not prescribed, because length is to be determined by the Contractor, it is impossible to state the mass of units except as the mass per unit of length (rule A6).

The criterion for allocating mark numbers to different units is given by rule A2. Its effect is that units which require different moulds because they have different dimensions will appear in different items. It is helpful to use main mark numbers (such as 61, 62) to distinguish different mould shapes and subdivisions of these (such as 61/a, 61/b, 61/c) to distinguish minor variations affecting length, position of holes or reinforcing details.

Rule D3 is given in order to exclude from class H a major concrete component such as a complete railway bridge deck which is cast adjacent to its final position and then moved into its final position during a line occupation. It would be wrong to measure such concrete work as precast since the work to be undertaken by the Contractor has all the characteristics of in situ concrete. Where there are very large precast concrete beams to be cast along the line of a bridge and then rolled and launched over each span, these would be measured as precast units because multiple use of formwork is involved as well as casting of the beams in other than their final position.

The necessity for giving separate items for each different shape of precast unit can sometimes give rise to more bill items than used to be traditional. For example, a breakwater constructed from raked and keyed precast concrete blocks may contain many different shapes. However, different shapes have different manufacturing costs and the number of different bill items reflects the complexity of the work. If a bill for precast concrete has a small number of items with large numbers of units against each this generally indicates lower manufacturing cost than does a bill with a large number of items with small numbers of units against each item.

*Schedule of changes in CESMM2*

1. Mass per metre of units measured linearly is to be stated in item descriptions.
2. Weir blocks laid to precise levels are not separately identified.

BRIDGES

| Number | Item description | Unit | Quantity | Rate | Amount £ | p |
|---|---|---|---|---|---|---|
| | PRECAST CONCRETE. | | | | | |
| | Ordinary prescribed mix concrete grade C40 as Specification clause F2/36. | | | | | |
| | Strefford Bridge Deck. | | | | | |
| H113.1 | Secondary beams inverted tee 180 x 400 mm length 4.25 m mass 500 kg-1t mark B1. | nr | 32 | | | |
| H113.2 | Secondary end beams rectangular 200 x 450 mm length 4.25 m mass 500 kg-1t mark B2. | nr | 4 | | | |
| H268 | Prestressed pre-tensioned main beams I section 800 x 1400 mm length 23.1 m mass 31 t mark B3 prestressing as drawing 136/21 and Specification clauses H2/21-27. | nr | 2 | | | |
| H523 | Service duct cover slabs thickness 100 mm area 1 - 4 m2 mass 500 kg-1t mark S1. | nr | 28 | | | |
| H810 | Parapet coping units filleted rectangular 250 x 220 mm cross-sectional area 0.05 m2 mass 120 kg/m mark C1. | m | 124 | | | |
| | | | PAGE TOTAL | | | |

# *Classes I–L: Pipework*

The variety of the components of civil engineering contracts which is covered by the general name of pipework is so large that the rules for measurement of pipes and associated work occupy four classes of the Work Classification. As a result the rules which these classes contain are closely interrelated and it is advisable for users of the CESMM to consider them as one composite class.

The rules in class I produce the bill items for runs of pipes, those in class J for pipe fittings and valves and those in class K for manholes and work associated with cross-country pipe laying and sewer work. Class L deals with work related to laying the pipes themselves, and extra cost items for trenching, bedding, haunching, wrapping and pipe supports. For CESMM2 the pipework classes were reviewed in considerable depth. A large number of changes of varying significance were made. A new class dealing with renovation of sewers was introduced as class Y. This class is described and explained later in this book.

## *Class I: Pipework—Pipes*

Class I provides the rules for measuring pipes and pipe runs. The items include excavating and backfilling trenches where pipes are laid in trenches and only the provision, jointing and laying of pipes when pipes are not in trenches.

The CESMM requires the components of pipelines to be described in much more detail than could be accommodated in the 512 possible classifications generated by the table in class I. To achieve this the first two divisions of the classification table are overriden by rule A2 and do not therefore affect item descriptions. Rule A2 requires the actual nominal bore (internal diameter) of the pipes to be stated and consequently, by virtue of paragraph 3.9, requires separate items to be given for each different nominal bore of pipe. Similarly rule A2 also produces separate items for pipes of different materials, with different types of joint, and with different lining requirements. The first and second division lists in the classification table only provide a broad categorization to maintain the numbering system for its uses outside the Bill of Quantities itself. This is the arrangement referred to in CESMM2 in the new paragraph 3.10.

Rule A1 is a very important rule. It requires all items for runs of pipe to state the location or type of pipework covered. It is easy to state location in broad terms in order to comply with the letter of the rule, but judgement is required to make the statement and the resulting separate items most useful for estimating and valuing variations. This judgement requires the work to be divided into lengths of pipe laying and excavation of trenches which are likely to involve work with particular cost characteristics—the type of judgement which pervades compilation of bills of quantities using the CESMM. This means giving separately identified and located items for pipework laid across different types of terrain, such as along roads, through back gardens, between buildings which limit access and choice of plant and through ground which is known to be either easy or difficult to support during trenching.

The cost of pipe laying is affected much more by the terrain and ground through which the work has to pass than by the precise nominal width and depth of trenches. Rule A1 requires separation of items to distinguish lengths where any part of the pipe laying operation as a whole is likely to be conducted differently or at a different pace due to favourable or unfavourable conditions in particular locations.

Rule A1 refers to location or type of pipework in order to provide for pipework the eventual use of which is more relevant to cost than its location. An example of this might be drainage pipework in new road construction. Here it would be more helpful to group together gully connections as a type of pipework than to divide all the pipework into lengths of the road itself designated by chainage.

The means of dealing with the depth of pipe trenches in the CESMM differs from other and previous methods of measurement and therefore justifies some explanation and illustration. In

CESMM2 the arrangements are unaltered except for a revision of the depth ranges to make them narrower at greater depths. Below 1·5 m, the depth range width is now consistently 0·5 m. The average depth is not stated in item descriptions; only the depth range or zone within which the pipe is laid is given. If, for example, a pipe were to be measured running from manhole A through manholes B and C to manhole D for a length of 385 m, the total length would be divided between as many items as were necessary to indicate the zones of depth within which the pipe was to be laid. It is not necessary to locate the ends of lengths which fall within the various depth ranges, provided the ends of a run of pipe are identified as, for example, between the manholes A and D. This should not be carried to excess by locating pipework as 'between Liverpool and Manchester', as this would not enable the tenderer to identify particular runs of pipe on the Drawings as he must be enabled to do from an item description or heading drawn up in accordance with rule A1.

This method of classifying trench depth is shown in Fig. 20. It is coupled with the procedure for identifying pipe runs by location provided by rule A1. It produces a small group of items which covers a run of pipe whose cost-significant factors to do with location do not change within its length. This group of items comprises separate items for the part of the length which falls within each depth zone given in the third division. The pipe and location descriptions are often put into a heading, followed by perhaps three items for the lengths which fall within three depth zones.

It is easy to take off quantities for bill preparation by this method. A straight edge is marked off to show the depth zones on the vertical scale of the longitudinal section drawings. The edge is moved across the drawing which is then marked where the depth from the Commencing Surface (which is usually the Original Surface) to the pipe invert crosses a depth zone mark. The lengths between marks are measured along the pipe to produce the quantities which go into the bill. Remeasurement for the final account is carried out in the same way.

It is not necessary to work out average depths or to adjust rates to account for small variations in average depth. The disadvantage of the method is that it does not give average depths from which estimators can calculate excavation and backfilling volumes. This is not

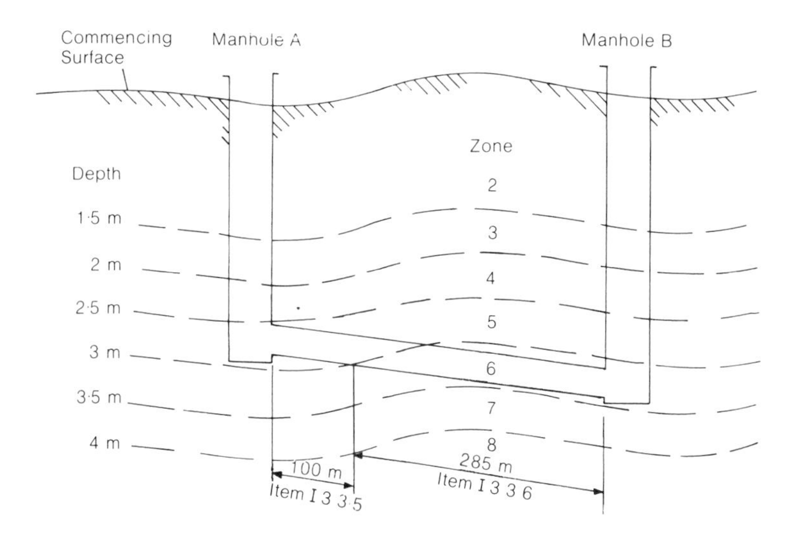


Fig. 20. Measurement of pipe trench depths. The depths of pipe trenches are given in zones (measured to pipe invert level). Thus any variation in trench depth at remeasurement shows itself as a change to the lengths of pipe which occur in each depth zone

a serious disadvantage. Calculation of these volumes was always approximate due to uncertainty about trench widths. The CESMM is likely to give separate items which individually cover work which is consistent as regards whether the trench will be battered or straight sided and supported, and more consistent as regards the pace of progress the plant/labour team will achieve.

The estimated plant and labour cost per metre of excavating and backfilling a trench is given by average depth (m) × average width (m) × gang cost (£/h) ÷ average rate of working ($m^3$/h). Using other methods of measurement the first of these factors was known precisely; the second and fourth were uncertain. Using the CESMM the first is known slightly less precisely; the second and fourth can be judged more precisely. In total, provided that rule A1 is applied thoroughly, the accuracy of estimating is not impaired by the change to simpler measurement.

The average depth to invert can in any case be determined with reasonable precision. If the items for one pipe run contain lengths in three zones or more the average depth of the intermediate zones is never likely to be more than about 150 mm from the mid-point of those zones. The average depth for the shallowest zone will tend to be on the deeper side of the mid-point and for the deepest zone on the shallower side of the mid-point. Average trench widths were seldom predicted with any greater precision.

The method produces a realistic adjustment of price in the event of variations of depth occurring between tender and final account. Any changes in depth either increase or decrease the lengths measured within each depth zone. The Contractor is paid for the actual quantity which occurs within the depth zone at the original rate for each item. If the depths are reduced the quantities at the lower rates for shallower depths increase and the quantities at higher rates for deeper depths decrease. A fine adjustment of the total price for constructing the pipe run covered by the group of items is achieved without complex measurement or calculation and without adjustment of rates.

Most advances are bought at a little cost, and this procedure does not work in one particular situation. If, by chance, the entire length of one pipe run were to occur within one depth zone and the average depth were to change without any part of the length falling into a different zone, remeasurement would produce no change in the price although the cost might have changed. An extreme case would be if the depth of a run had been expected to vary between 3·0 m and 3·2 m but in the event it varied between 3·3 m and 3·5 m. The average depth would change from 3·1 m to 3·4 m, increasing the volume of excavation and backfill by 2 m$^3$ per metre if a battered trench were involved. The converse is equally likely to occur, but occasionally where the depth changes do not mutually compensate it will be necessary to adjust the rate.

The remeasurement of a pipe run subject to changes in depth is shown in Fig. 21. This is simplified to make the point by assuming straight lines for the Commencing Surface and for the pipe runs. It shows how remeasurement of the lengths in each depth zone leads to adjustment of the total price for the pipe run. If the rates for each zone allow for the cost of excavation and backfilling appropriate to

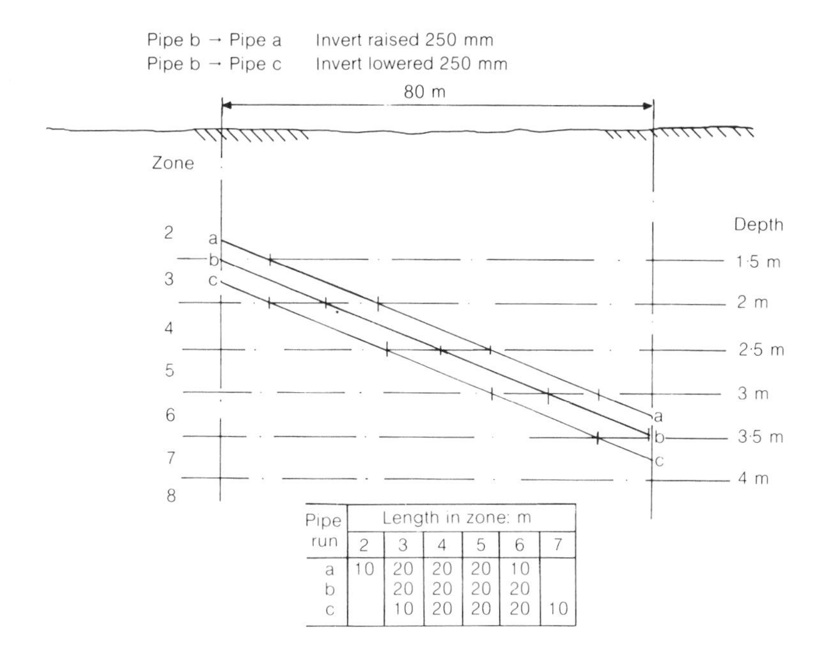


| Pipe run | Length in zone: m | | | | | |
|---|---|---|---|---|---|---|
| | 2 | 3 | 4 | 5 | 6 | 7 |
| a | 10 | 20 | 20 | 20 | 10 | |
| b | | 20 | 20 | 20 | 20 | |
| c | | 10 | 20 | 20 | 20 | 10 |

Fig. 21. Simplified example showing how remeasurement of a pipe run in depth zones deals with variations in pipe trench depths

the average depth for the zone, remeasurement is bound to produce adjustment which corresponds very closely to the actual change in volumes. Estimators who are sceptical are recommended to check this using Fig. 21 and assumed costs per linear metre for excavation and backfilling which are proportional to the cross-sectional areas expected at each zone.

Note that depths are measured from the Commencing Surface to the pipe invert (rule D3). Allowance for the volume of excavation below that level should be made in rates for all depth zones. This should not include further excavation for beds to pipes as this is included in the rates against the items for beds in class L (rule C1). Lengths of pipes are measured along their centre lines (rule M3).

This means that the plan length is less than the true length for steeply inclined pipes. The true length should be used for measurement when it is noticeably different from the plan length.

Rule A5 requires item descriptions to state where more than one pipe is laid in one trench. This produces separate items for each pipe under the same locational heading. Each pipe will be classified according to its own depth. The addition of 'in shared trench' to a heading will make the situation clear to tenderers. Where the sequence of items adopted prevents separate runs in shared trenches being billed consecutively the text 'in shared trench FG' added to both item descriptions will serve the same purpose.

Rule M3 states the measurement convention that the lengths measured for pipelines in trenches shall include the lengths occupied by fittings and valves. Fittings and valves on pipes in trenches are effectively priced extra over the pipe runs in which they occur. This is a convenient and effective convention because it means that prices for fittings and valves need not allow for excavation and backfilling unless a major widening of a trench is required. No measurement of the lengths occupied by the various fittings and lengths of pipes within one run is required. This arrangement is changed in CESMM2 for measurement of fittings and valves on pipes not in trenches. It has eliminated the disadvantages which attended the former method, particularly in the measurement of complex pipework such as pipework in water treatment works, particularly if detailed drawings were not given. If the fittings and valves come so fast and furiously that there is very little plain pipe between them, and the short lengths of plain pipe all have a different nominal bore, it can be very difficult to decide what lengths of what non-existent plain pipe would have been there if they had not been displaced by fittings and valves. The last sentence of rule M3 now makes this unnecessary by requiring fittings and valves on pipes not in trenches to be measured full value.

Class I does not give a rule for deciding on the point from which to measure pipe lengths when pipes in trenches of different bore or material issue from opposite sides of a valve or from the branches of a junction. However, it would be difficult to mount an argument for doing anything other than taking the measurement of each pipe from the centre point of the fitting or valve.

Rules M4, D1 and D2 govern the itemization of and descriptions for pipes not in trenches (items I * *1). Their effect is that this designation is used for any pipes for which the items in class I are not intended to include excavation and backfilling of trenches. Some of the alternative ways of getting pipes into position, such as laying pipes in heading, could be a matter for the Contractor's discretion. Designation of pipes as 'not in trench' is therefore made conditional on such construction being expressly required (rule D2). Where it is expressly required, appropriate items are given in classes K and L. These items are mainly those in class L for work associated with headings, thrust boring and supporting pipes above the ground. As these items are given separately, the prices set against the items for pipes not in trenches in class I remain only to cover the cost of the provision of the pipes themselves, of jointing and setting in position (rule M4).

*Class J: Pipework—Fittings and valves*

The measurement rules in the CESMM for fittings and valves to pipework are perhaps the simplest in the Work Classification. All items for fittings and valves are given a comprehensive description and the quantities are the numbers of each type required.

Rule A1 requires the same details to be given in item descriptions for pipe fittings such as bends and junctions as are required in class I for pipes themselves—nominal bores, materials, jointing and lining details. Lengths and angles of bends, junctions and branches are not required to be stated. As in class I, the classification table is overridden by this rule. In class J it is the first and third divisions which are demoted instead of the first and second. Rule A5 establishes similar requirements for the descriptions of items for valves and penstocks.

The list of fittings and valves given in the second division of the classification table is not intended to be complete. It allocates classification numbers to only the most commonly encountered components. Many other types of fittings and valves appear in bills of quantities, and, if coded, are given item numbers J * 9 *.

In the second edition of the CESMM a new item 'straight specials' has been introduced (J 1–7 8 *). Rule D2 defines the fittings which are classed as straight specials. They are non-standard lengths of pipe which are expressly required to be provided. They would not be measured where non-standard lengths are necessitated only by the layout of the work. Whether straight specials are fabricated or cut on Site may either be left to the Contractor's discretion or stated in accordance with rule A1. In either case the item will cover all the costs of fabrication or cutting by virtue of rule C3. Where a Contractor introduces short lengths of pipes for purposes of his own, they will not be classed as straight specials but will be covered by the main items in class I.

Rule A2 introduces a special rule for cast or spun iron pipe fittings of nominal bore exceeding 300 mm and for all pipe fittings made of steel. These fittings are particularly costly and more detailed description is necessitated. The 'principal dimensions' referred to in this rule are those which will identify the fittings in a supplier's catalogue. To state the effective length, nominal bore and angle of a bend would be sufficient to comply with this rule.

Rule D1 is a convenient and sensible criterion to apply to the classification of cast iron pipe fittings which have one flange at a nominal bore of less than 300 mm and one greater, and to apply to the statement of the nominal bore of junctions, branches and tapers in complying with rule A1.

Rule A4 requires fittings to pipework not in trenches to be measured separately. In accordance with the rules in class I, this convention will distinguish those fittings which are measured full value from those which are effectively measured extra over.

Estimators should note that the CESMM does not make it a rule that any special plant or labour cost due to the positions of fittings and valves in the Works should be reflected in item descriptions. Such additional information or categorization will be given only where bill compilers judge it helpful in their application of paragraph 5.10.

### *Class K: Pipework—Manholes and pipework ancillaries*

The rules of measurement in class K cover a variety of work associated mainly with drainage and cross-country pipe laying. Manholes and other chambers are referred to in the classification table at K 1–2 * *. In CESMM2, the ranges of depths of manholes and other chambers have been revised and narrowed consistently with the depth ranges for pipe trenches in class I.

The components of manholes, other chambers and gullies are not required to be measured in detail in the Bill of Quantities. They are identified by a type or mark number which refers to details in the Drawings and the Specification. These details do not need to be specific as regards arrangements of inlets and outlets or heights of access shafts to comply with rule A1. Different configurations of manholes and other chambers such as catchpits and gullies are required to be given in separate items, with separate type or mark numbers stated. The procedure set down in paragraph 3.11 has the effect that the work covered by rules C1 and C2 is also deemed to be included in the items for manholes and other chambers.

Separate items are not given for linings or surrounds to manholes and other chambers. It is consequently necessary to ensure that chambers which are the same except for having different linings or surrounds are given separate type or mark numbers. Excavation and backfilling for chambers is not measured separately except for the extra cost items given in class L. Excavation and backfilling of working space around chambers is obviously also not measured. Estimators would normally assume that chamber cover levels are close to the Commencing Surface for excavation. It would be a proper application of paragraph 5.10 for bill compilers to draw attention to instances where this is not the case, and to any other special cost-significant aspect of particular chambers.

Manholes with backdrops are given separately in items K 1 * *. As a backdrop is consequently regarded as part of a manhole, it is not necessary to identify a short length of deeper trench approaching the manhole when giving items for the pipes comprising a backdrop which is built up from pipe lengths and fittings and not manufactured as an integral part of a precast concrete manhole. The fittings included in backdrops are not included in the quantities of fittings measured in class J (rule C4 of class K). Rule M3 of class I in CESMM2 is now consistent with rule C4 of class K as regards the inclusion of backdrop pipework in the manhole items in class K.

Other changes in CESMM2 relating to manholes and other chambers are the references to drawpits and the elimination of catchpits (now one of the category 'other chambers'). The depths of chambers can now be measured to the invert level if it is below the top of the base.

A note at the foot of page 51 of CESMM2 draws attention to the option of measuring manholes and other chambers in full detail using the other classes of the CESMM where the complexity, size or any other peculiarity of the work warrants so doing.

French and rubble drains are another special type of work which is measured in a complicated way partly as a result of the general rules having been made simple. Such a drain which has a pipe in the bottom will appear among the items in class I, separately identified to comply with rule A5 of class I. The rate against this item should allow at least for excavation of the trench, providing and laying the pipe and disposing of excavated material. It should not allow for backfilling with porous material as separate items measured in cubic metres are given for that in class K (items K 4 1–2 0). Rule M2 of class I also draws attention to this. Where a drain is formed by backfilling a trench with porous material and no pipe is provided, the excavation of the trench and disposal of excavated material is covered by the items for trenches for unpiped rubble drains (items K 4 3 *). Backfilling of such drains is covered by the same items as backfilling of piped drains (items K 4 1–2 0).

Class K includes the items for measurement of ducts and metal culverts at K 5 * *. These rules and the related rules in classes I and L were given a general overhaul in the production of CESMM2. They are not now intended to operate differently from the CESMM first edition but to be more coherent and clearer for the user of the CESMM. The basic principle, easier to follow perhaps than the detailed rules themselves, is that the measurement of ducts and metal culverts is in almost every respect the same as the measurement of pipes. This simple principle needs rules M3, M4, M6, M9, D5, C5 and A5 of class K and others in other classes to give it effect. Between them, they extend the application of the method of measurement for pipework to ducts and metal culverts and also establish that crossings, reinstatement and other pipework ancillaries should be treated identically whether they are associated with pipework or with ducts and metal culverts. Rule D7 defines the dimension used in the third division of classification for ducts and metal culverts in place of the nominal bore of ordinary pipes. It is illustrated in Fig. 22.

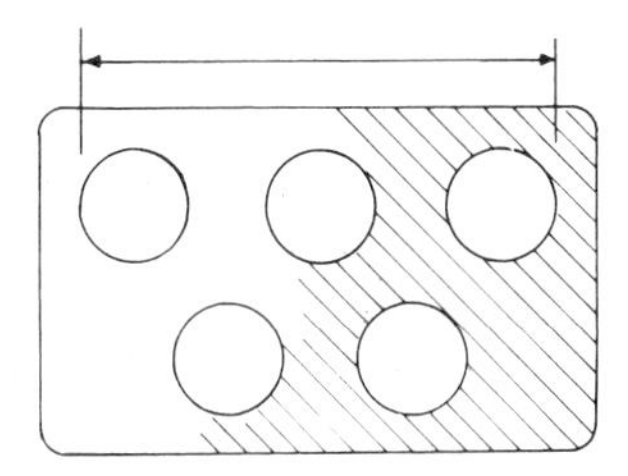

Fig. 22. Dimension used in calculating nominal trench widths for multiple bore ducts and twinned pipes. This is described in rules D1 and D7 of class K and A3, M2 and D1 of class L. The dimension is referred to as the 'distance between the inside faces of the outer pipe walls'

The rules governing the measurement of ditches are given in the classification table at K 4–7 *. The CESMM does not require the precise dimensions of ditches themselves to be given. Ditches are measured linearly with the areas of their cross-sections given within fairly narrow ranges. If a ditch is to be lined the item description must state the lining materials to be used and the thickness of the lining, and other relevant particulars (rule A4). Trenches for pipes or cables to be laid by persons other than the Contractor are included in the CESMM2 at K 4 8 *.

The measurement of crossings and reinstatement is governed by the classification table at K 6–7 * *. The items given are brief and do not distinguish work according to the type of pipe in the trench or the depth at which it is laid. The second edition of the CESMM provides items for crossings of existing services within pipe trenches. Coverage rules throughout CESMM2 require items for excavation to include removal of existing services in the ground which are dead and can just be stripped out. In this class, the CESMM now provides items for working around or reinstating existing live services which have to be maintained when encountered in the course of trenching for pipes and ducts. The items are given in the classification at K 6 7–8 1–4. Field drains are not included as crossings of sewers or drains as they are covered specifically by item K 8 1 0. The items for crossings of services are measured by number of crossings, no distinction being made between those at right angles to the line of the trench and those almost parallel to it. Only crossings of underground services are measured. Crossings outside the nominal excavated area should not be measured.

The list of nominal bore ranges given in the classification table in the third division at K 6–7 * 1–4 is different from that used in classes I, J and L. This is because the costs of crossings and reinstatement are not dependent on the precise nominal bore of the pipes and the items can be grouped into relatively wide ranges. The bore used for classification in the third division is, in any case, the aggregate bore of the pipes in the trench (rule D7). Rule D6 in CESMM2 utilizes the new procedure for dealing with work affected by water (paragraph 5.20) to define the length of pipe trenches affected by crossings of rivers, streams and canals.

CESMM2 adds an item for stripping topsoil from the easement for pipelines and reinstating (K 7 6 0). It is governed by rules M8, C9, A8 and A9. Where this item is used, it should be considered as covering work across the full width of the easement including the nominal trench width. This means, amongst other things, that no other reinstatement is measured for the lengths of trench affected and that the Commencing Surface for the trench excavation itself becomes the underside of topsoil strip.

The item descriptions for connections of new pipework to existing work (K 8 5–6 *) should give information in accordance with rule A11 in order to separate different types of connection. The items are taken to cover the cost of all associated work such as maintaining flows and altering benching in manholes. It is consequently important to state requirements for such work clearly.

### *Class L: Pipework—Supports and protection ancillaries to laying and excavation*

Class L provides the rules for measurement of some of the operations involved in excavation and backfilling of pipe trenches which give rise to extra cost (the classification table at L 1 * *), for pipe laying in headings, by pipe jacking and by thrust boring (the classification table at L 2 * *) and for supports and protection to pipework (the classification table at L 3–8 * *).

The items for extras to excavation and backfilling cover excavation and backfilling which are more costly than ordinary soft excavation within the depth of a trench and replacement of this same material after pipe laying. The volumes involved are measured in addition to the lengths of pipelines in trenches given by the items in class I. Consequently, they should be priced only to cover the extra costs of dealing with the materials identified: excavation of rock, mass concrete, reinforced concrete and other artificial hard material, and backfilling with material other than that excavated.

The nature of special backfilling material must be stated. Selected excavated material is not regarded as special backfilling (rule M2 of class I). Items for backfilling are divided into those covering work above the Final Surface (L 1 * 5–6) and those covering work below the Final Surface (L 1 * 7–8). It may seem odd at first that any excavation and backfilling should take place below the Final Surface if that surface is truly final. However, the Final Surface is 'the surface indicated on the Drawings to which excavation is to be carried out' (paragraph 1.11). Therefore excavation below the Final Surface is required if soft spots or unsuitable material are found in the bottom of a pipe trench which has already been excavated to the levels indicated on the Drawings.

Rules M4 and D1 provide the measurement convention for calculating quantities of extras to trench excavation and backfilling. A convention for trench width is given which is used in calculating the actual volumes irrespective of the actual width of material. This means that the resulting quantity is notional to the extent that the prices in tenders must allow for the general undermeasurement of the actual volume which will result. The paid volume will be based on an assumed vertical-sided trench, whereas the prices will have to allow for the cost of additional excavation and backfill in any special materials where trench sides are battered. This consideration does not apply to the quantities of excavation and backfilling of soft spots below the Final Surface as these would normally be close to or sometimes less than the assumed width. The width is taken as 500 mm greater than the nominal bore of single pipes not exceeding 1 m nominal bore and as 750 mm greater when the nominal bore exceeds 1 m. Where two or more pipes are laid at the same level the nominal width is greater by the same amount than the total width occupied by the bores of the pipes. Where pipes are laid at different levels in one trench the nominal width should be assessed separately for the vertical distance between pipes.

The nominal trench width is multiplied by the average depth and average length of material removed or backfilled to give the measurement of volume. The average dimensions in this case are the average depth across the nominal width of the trench and the average length in the depth of the trench. In other words, the area to be multiplied by the nominal width is the average of the areas of the special materials which would appear on longitudinal sections drawn for the trenches. This will rarely differ from the area of the longitudinal section at the trench centre line. The same type of calculation is required for dealing with special materials in excavation for manholes and other chambers. Rule M5 states the rule which is applied. This rule differs from its predecessor in the first edition of the CESMM. It must be read with paragraph 5.18 to give it proper effect.

A new rule in this class governs the measurement of boulders and small volumes of hard material encountered when excavating pipe trenches or when installing pipes underground by special methods. This is rule M8 which is in the same terms as rule M8 in class E.

The items which have been discussed for extras to excavation and backfilling are the sort of items which were traditionally designated as provisional quantities. Under the fifth edition of the Conditions of Contract this designation no longer has any function and the compiler of the bill is required to forecast the likely quantities as well as he can. This point is considered in more detail in relation to paragraph 5.17 (see page 57).

The item descriptions for headings, pipe jacking and thrust boring are required to state location so that they can be cross-referenced to the items in class I for providing the pipes concerned (rule A1). These items, like the associated items in class I, are given only when the work is expressly required (rule M9). Thrust boring and pipe jacking require plant and Temporary Works which are likely to be covered by Method-Related Charges. If the pits for thrust boring and pipe jacking are specified requirements within the terms of rule D1 of class A the general items will contain items for them under that heading (rule M10). If the arrangements for pits are at the Contractor's discretion they will not be classed as specified requirements and tenderers will be free to insert Method-Related Charges for them if they so wish. Note the effect of rules C3 in class L and D1 of class I as worded in CESMM2. It is unequivocally that the cost of provision, laying and jointing of the pipes installed by special methods is covered by the items in class I (I * * 1). Only the additional cost due to the special pipelaying methods is covered by the items in class L (L 2 1–3 *).

Measurement of supports and protection is made in items which identify the materials to be used and state the range of nominal bore of the pipes to be treated. Paragraph 5.14 is brought into effect when only one bore size is included so that the actual nominal bore is then stated. The cross-sectional dimensions of beds, haunches and surrounds other than bed depths are not required to be stated, on the assumption that details will be given on the Drawings.

Rule C1 of class L makes it clear that, as separate items are not given for excavation for work such as beds to pipes, this should be allowed for in the items concerned. Note that this will be normal soft excavation, as extras will be measured for hard materials, and that the prices do not distinguish the depth at which beds are to be excavated or constructed. An effect of rules D2 and A2 is that where beds and haunches or beds and surrounds are made of the same material they are combined as one item as shown in the example bill pages.

The items for concrete stools and thrust blocks are very simple. They are enumerated and identified by item descriptions which state the specification of the concrete, whether it is reinforced and the range of concrete volume within which each falls. Rule C2 establishes that the items include formwork, reinforcement, joints and finishes.

Items for pipe supports are classified by height as defined in rule D5. Their principal dimensions and materials must be stated to comply with rule A6. Occasionally concrete pipe stools and metal pipe supports will be large enough to justify measurement as concrete or metal structures in their own right although their function is clearly only to support pipes. Bill compilers should not set the boundary too high as the rules given in class L are not appropriate to the measurement of substantial structures.

## *Schedule of changes in CESMM2*

*Class I*

1. Trench depth ranges are narrowed.
2. Lengths of pipes not in trenches exclude lengths occupied by fittings and values.

*Class J*

3. Items are included for expressly required straight special pipes.
4. More detailed description is required for some fittings.
5. Fittings and valves on pipework outside trenches are shown separately.

*Class K*

6. Depth ranges for manholes and other chambers are revised and narrowed.
7. Items are added for trenches to receive pipes or ducts laid by others.
8. Items are added for service crossings within pipe trenches.
9. Items are added for stripping topsoil from pipeline easements and reinstatement.
10. General overhaul of rules for measurement of ducts and metal culverts.

*Class L*

11. Minimum volume is stated for rock and other hard material encountered in pipe trenches.
12. Pipe jacking items are added.
13. Rules for beds, haunches and surrounds are amended and expanded.
14. Rules for stools, thrust blocks and pipe supports are amended and expanded.

| Number | Item description | Unit | Quantity | Rate | Amount £ | p |
|---|---|---|---|---|---|---|
| | PIPEWORK - PIPES. | | | | | |
| | Clay pipes to BS 65 with spigot and socket flexible joints nominal bore 225 mm in trenches. | | | | | |
| | Between manholes 7 and 11. | | | | | |
| I123.1 | Depth: 1.5 - 2 m. | m | 187 | | | |
| I124.1 | Depth: 2 - 2.5 m. | m | 291 | | | |
| I125 | Depth: 2.5 - 3 m. | m | 102 | | | |
| | Between manholes 27 and 31. | | | | | |
| I123.2 | Depth: 1.5 - 2 m. | m | 113 | | | |
| I124.2 | Depth: 2 - 2.5 m. | m | 202 | | | |
| | Prestressed concrete pipes to BS 5911 (class M) with ogee joints nominal bore 375 mm. | | | | | |
| I231 | Between manholes 18 and 19 installed by thrust boring under Central Wales line railway embankment. | m | 98 | | | |
| | Between manholes 19 and 23 in trenches. | | | | | |
| | Commencing Surface underside of topsoil. | | | | | |
| | (Stripping measured separately in item number K760.1). | | | | | |
| I234 | Depth: 2 - 2.5 m. | m | 127 | | | |
| I235 | Depth: 2.5 - 3 m. | m | 78 | | | |
| I238 | Depth: 4.5 - 5 m. | m | 21 | | | |
| | | | PAGE TOTAL | | | |

SITE PIPELINES

| Number | Item description | Unit | Quantity | Rate | Amount £ | p |
|---|---|---|---|---|---|---|
| | PIPEWORK - PIPES. | | | | | |
| | Prestressed concrete pipes to BS 5911 (class M) with ogee joints nominal bore 375 mm. | | | | | |
| | Manhole 23 to primary tank. | | | | | |
| I233 | Depth 1.5 - 2m; in trench laid alongside second pipe. | m | 27 | | | |
| | Cast or spun iron pipes to BS 4622 (class 3) with Forsoam joints nominal bore 700 mm. | | | | | |
| I441 | Feed pipe supported above the ground. | m | 16 | | | |
| I443 | In trench depth 1.5 - 2 m Commencing Surface underside of main tank slab; draw off pipe main tank collector to valve chamber. | m | 12 | | | |
| I445 | In trench D depth 2.5 - 3 m bypass pressure main valve chamber to circulation unit. | m | 33 | | | |
| I446 | In trench D depth 3 - 3.5 m service pressure main valve chamber to circulation unit. | m | 33 | | | |
| | Plastics pipes to BS 3505 with compression joints nominal bore 50 mm washwater mains. | | | | | |
| | In trenches. | | | | | |
| I612 | Depth: not exceeding 1.5 m. | m | 98 | | | |
| I613 | Depth: 1.5 - 2 m. | m | 150 | | | |
| I614 | Depth: 2 - 3 m. | m | 128 | | | |
| | | | PAGE TOTAL | | | |

| Number | Item description | Unit | Quantity | Rate | Amount £ | p |
|---|---|---|---|---|---|---|
| | PIPEWORK - FITTINGS AND VALVES. | | | | | |
| | Clay pipe fittings to BS 65 with spigot and socket flexible joints; nominal bore 225 mm. | | | | | |
| J112 | Bends. | nr | 385 | | | |
| J122 | Junctions and branches. | nr | 127 | | | |
| J132 | Tapers. | nr | 38 | | | |
| | Prestressed concrete pipe fittings to BS 5911 (class M) with ogee joints. | | | | | |
| J212 | Bends; nominal bore 225 mm. | nr | 41 | | | |
| J213.1 | Bends; nominal bore 375 mm. | nr | 141 | | | |
| J213.2 | Bends; nominal bore 450 mm. | nr | 19 | | | |
| J214 | Bends; nominal bore 825 mm. | nr | 47 | | | |
| J222 | Junctions and branches; nominal bore 225 mm. | nr | 27 | | | |
| J223.1 | Junctions and branches; nominal bore 375 mm. | nr | 76 | | | |
| J223.2 | Junctions and branches; nominal bore 450 mm. | nr | 8 | | | |
| J224 | Junctions and branches; nominal bore 825 mm. | nr | 23 | | | |
| J232 | Tapers; nominal bore 225 mm. | nr | 16 | | | |
| J233 | Tapers: nominal bore 450 mm. | nr | 16 | | | |
| J234 | Tapers; nominal bore 825 mm. | nr | 4 | | | |
| | | | PAGE TOTAL | | | |

SITE PIPELINES

| Number | Item description | Unit | Quantity | Rate | Amount £ | p |
|---|---|---|---|---|---|---|
| | PIPEWORK - FITTING AND VALVES. | | | | | |
| | Cast or spun iron pipe fittings to BS 4622 (class 3) with Grippon joints. | | | | | |
| J411 | Bends; nominal bore 150 mm. | nr | 27 | | | |
| J413.1 | Bends; nominal bore 400mm 22.5° effective length 220 mm; not in trenches. | nr | 1 | | | |
| J413.2 | Bends; nominal bore 400 mm 45° effective length 400 mm. | nr | 6 | | | |
| J413.3 | Bends; nominal bore 500 mm vertical 90° effective length 840 mm. | nr | 3 | | | |
| J453 | Adaptors; nominal bore 400 mm Grippon to flange effective length 160 mm. | nr | 6 | | | |
| J482.1 | Straight specials nominal bore 225 mm effective length 1220 mm. | nr | 1 | | | |
| J482.2 | Straight specials nominal bore 225 mm effective length 1500 mm; with puddle flange. | nr | 1 | | | |
| J483 | Straight specials nominal bore 400 mm effective length 900 mm; not in trenches. | nr | 3 | | | |
| J493 | Blank flanges; nominal bore 400 mm. | nr | 3 | | | |
| | Cast iron gate valves hand operated to BS 3464 (type a) with extension spindles. | | | | | |
| J813.1 | 400 mm nominal bore with T key. | nr | 4 | | | |
| J813.2 | 400 mm nominal bore with hand wheel. | nr | 2 | | | |
| J813.3 | 500 mm nominal bore with hand wheel. | nr | 3 | | | |
| | | | | PAGE TOTAL | | |

SITE PIPELINES

| Number | Item description | Unit | Quantity | Rate | Amount £ | p |
|---|---|---|---|---|---|---|
| | PIPEWORK - FITTINGS AND VALVES. | | | | | |
| | Cast iron valves as Holdwater Industries catalogue numbers stated or similar approved. | | | | | |
| J884.1 | Penstocks nominal bore 700 mm nr 731 with handwheel headstockand foot bracket; invert to handwheel distance 2050 mm. | nr | 6 | | | |
| J884.2 | Penstocks nominal bore 700 mm nr 740 with T key headstock and foot bracket; invert to T key distance 3100 mm. | nr | 9 | | | |
| J894 | Sludge draw off valves nominal bore 700 mm nr 780. | nr | 3 | | | |
| J895 | Double hinged flap valves nominal bore 1100 mm nr 890. | nr | 2 | | | |
| | | | PAGE TOTAL | | | |

CHAT MOSS PIPELINE

| Number | Item description | Unit | Quantity | Rate | Amount £ | p |
|---|---|---|---|---|---|---|
| | PIPEWORK - MANHOLES AND PIPEWORK ANCILLARIES. | | | | | |
| | Manholes. | | | | | |
| K111 | Brick depth not exceeding 1.5 m; type A1 with medium duty cast iron cover to BS 497 reference MB1-55. | nr | 7 | | | |
| K112.1 | Brick depth 1.5 - 2 m; type A1 with heavy duty triangular cast iron cover to BS 497 reference MA-T. | nr | 7 | | | |
| K112.2 | Brick depth 1.5 - 2 m; type A2 with medium duty cast iron cover to BS 497 reference MB2-55. | nr | 4 | | | |
| K122.1 | Brick with backdrop depth 1.5 - 2 m; type A10 with cast iron cover to BS 497 reference MC2-60/45. | nr | 5 | | | |
| K122.2 | Brick with backdrop depth 1.5 - 2 m; type A20 with cast iron cover to BS 497 reference MC1-60/60. | nr | 3 | | | |
| K152 | Precast concrete depth 1.5 - 2 m; type C1 with triangular cast iron cover to BS 497 reference MA-T. | nr | 33 | | | |
| K211 | Catch pits brick depth not exceeding 1.5 m; type K1 with light duty cast iron cover to BS 497 reference C6-24/24. | nr | 6 | | | |
| K360 | Gullies precast concrete trapped; with medium duty straight bar gulley grating to BS 497 reference E12-12. | nr | 15 | | | |
| K410 | Filling of French and rubble drains with graded material; type C3 granular material. | $m^3$ | 310 | | | |
| K433 | Trenches for unpiped rubble drains cross-sectional area 0.5 - 0.75 m2. | m | 54 | | | |
| | | | PAGE TOTAL | | | |

CHAT MOSS PIPELINE

| Number | Item description | Unit | Quantity | Rate | Amount £ | p |
|---|---|---|---|---|---|---|
| | PIPEWORK - MANHOLES AND PIPEWORK ANCILLARIES. | | | | | |
| K434 | Trenches for unpiped rubble drains cross-sectional area 0.75 - 1 m2. | m | 89 | | | |
| K453 | Rectangular section ditches cross-sectional area 0.5 - 0.75 m2; lined with 1000 gauge melded fibre mat. | m | 28 | | | |
| | Ducts and metal culverts. | | | | | |
| K542 | Unglazed vitrified clay 4 way 100 mm nominal bore cable duct loose jointed in trenches depth not exceeding 1.5 m beneath roads. | m | 14 | | | |
| K543 | Unglazed vitrified clay 4 way 100 mm nominal bore cable duct loose jointed in trenches depth 1.5 - 2 m beneath roads. | m | 25 | | | |
| K553 | Sectional corrugated metal culverts as Specification clause K56/3 nominal internal diameter 700 mm in trenches depth 1.5 - 2 m in outfall. | m | 63 | | | |
| | Crossings. | | | | | |
| K623.1 | River width 3 - 10 m pipe bore 900 - 1800 mm. | nr | 2 | | | |
| K623.2 | Canal width 3 - 10 m; pipe bore 300 - 900 mm. | nr | 4 | | | |
| K641 | Hedge; pipe bore not exceeding 300 mm. | nr | 30 | | | |
| K671 | Sewer or drain; pipe bore not exceeding 300 mm. | nr | 15 | | | |
| K672 | Sewer or drain; pipe bore 300 - 900 mm. | nr | 7 | | | |
| K682.1 | Gas main; pipe bore 300 - 900 mm. | nr | 2 | | | |
| | | | | PAGE TOTAL | | |

CHAT MOSS PIPELINE

| Number | Item description | Unit | Quantity | Rate | Amount £ | p |
|---|---|---|---|---|---|---|
| | PIPEWORK - MANHOLES AND PIPEWORK ANCILLARIES. | | | | | |
| | Crossings. | | | | | |
| K682.2 | Existing sludge draw off pipe line: pipe bore 300 - 900 mm. | nr | 2 | | | |
| K683 | Underground high voltage electric cable: pipe bore 900 - 1800 mm. | nr | 1 | | | |
| | Reinstatement. | | | | | |
| K711 | Breaking up and temporary reinstatement of roads flexible road construction maximum depth 75 mm with 250 mm sub-base; pipe bore not exceeding 300 mm. | m | 387 | | | |
| K712.1 | Breaking up and temporary reinstatement of roads reinforced concrete slab depth 200 mm with 250 mm type 1 sub-base; pipe bore 300 - 900 mm. | m | 284 | | | |
| K712.2 | Breaking up and temporary reinstatement of roads flexible road construction maximum depth 75 mm with 250 mm type 1 sub-base; pipe bore 300 - 900 mm. | m | 783 | | | |
| K741 | Breaking up temporary and permanent reinstatement of footpaths precast concrete paving flags maximum depth 75 mm with 50 mm sand bed; pipe bore not exceeding 300 mm. | m | 54 | | | |
| K760 | Strip topsoil from easement and reinstate in cultivated land; minimum width 20 m. | m | 3726 | | | |
| | | | | PAGE TOTAL | | |

CHAT MOSS PIPELINE

| Number | Item description | Unit | Quantity | Rate | Amount £ | p |
|---|---|---|---|---|---|---|
| | PIPEWORK - MANHOLES AND PIPEWORK ANCILLARIES. | | | | | |
| | Other pipework ancillaries. | | | | | |
| K810 | Reinstatement of field drains. | m | 247 | | | |
| K820 | Marker posts; 100 x 100 x 1500 mm hardwod set in concrete base as drawing 137/15 detail C. | nr | 47 | | | |
| K830 | Timber supports left in excavations. | m2 | 450 | | | |
| K852 | Connection of pipes to existing manholes; nominal bore 200 - 300 mm; foul water sewer including breaking into manhole, reforming benching and dealing with flows in accordance with detail 7 drawing C3/21. | nr | 1 | | | |
| | | | | PAGE TOTAL | | |

CHAT MOSS PIPELINE

| Number | Item description | Unit | Quantity | Rate | Amount £ | p |
|---|---|---|---|---|---|---|
| | PIPEWORK - SUPPORTS AND PROTECTION ANCILLARIES TO LAYING AND EXCAVATION. | | | | | |
| | Extras to excavation and backfilling in pipe trenches. | | | | | |
| L111 | Excavation of rock. | m3 | 460 | | | |
| L112 | Excavation of mass concrete. | m3 | 65 | | | |
| L113 | Excavation or reinforced concrete. | m3 | 85 | | | |
| L115 | Backfilling above the Final Surface with concrete; grade C10. | m3 | 2 | | | |
| L116 | Backfilling above the Final Surface with imported natural material other than rock or topsoil. | m3 | 310 | | | |
| L117 | Excavation of natural material below the Final Surface and backfilling with concrete: grade C10. | m3 | 260 | | | |
| L118.1 | Excavation of natural material below the Final Surface and backfilling with type C2 granular material. | m3 | 270 | | | |
| L118.2 | Excavation of natural material below the Final Surface and backfilling with type C3 granular material. | m3 | 230 | | | |
| | Extras to excavation and backfilling. | | | | | |
| L121 | In manholes and other chambers excavation of rock. | m3 | 27 | | | |
| L122 | In manholes and other chambers: excavation of mass concrete. | m3 | 4 | | | |
| L141 | In thrust boring: excavation of rock. | m3 | 127 | | | |
| | | | | PAGE TOTAL | | |

CHAT MOSS PIPELINE

| Number | Item description | Unit | Quantity | Rate | Amount £ | p |
|---|---|---|---|---|---|---|
| | PIPEWORK - SUPPORTS AND PROTECTION ANCILLARIES TO LAYING AND EXCAVATION. | | | | | |
| | Special pipe laying methods. | | | | | |
| L223 | Thrust boring pipe nominal bore 300 - 600 mm; manholes 18 to 19. | m | 98 | | | |
| | Beds depth 150 mm. | | | | | |
| L331 | Imported granular material type C2 pipe nominal bore not exceeding 200 mm. | m | 27 | | | |
| L332 | Imported granular material type C2 pipe nominal bore 200 - 300 mm. | m | 2483 | | | |
| L352 | Reinforced concrete grade C15 reinforced as drawing 137/19 detail B pipe nominal bore 200 - 300 mm. | m | 125 | | | |
| L353 | Reinforced concrete grade C15 reinforced as drawing 137/19 detail A pipe nominal bore 300 - 600 mm. | m | 90 | | | |
| | Haunches. | | | | | |
| L431 | Imported granular material type C2 pipe nominal bore not exceeding 200 mm; bed depth 200 mm. | m | 50 | | | |
| L432 | Imported granular material type C2 pipe nominal bore 200 - 300 mm; bed depth 200 mm. | m | 250 | | | |
| | Surrounds. | | | | | |
| L531.1 | Imported granular material type C2 pipe nominal bore not exceeding 200 mm; bed depth 200 mm. | m | 402 | | | |
| | | | PAGE TOTAL | | | |

CHAT MOSS PIPELINE

| Number | Item description | Unit | Quantity | Rate | Amount £ | p |
|---|---|---|---|---|---|---|
| | PIPEWORK - SUPPORTS AND PROTECTION ANCILLARIES TO LAYING AND EXCAVATION. | | | | | |
| | Surrounds. | | | | | |
| L531.2 | Imported granular material type C2 to two pipes, maximum distance between inside face of outer walls of pipes 600 mm; bed depth 200 mm. | m | 41 | | | |
| L601 | Wrapping and lagging pipe nominal bore not exceeding 200 mm with Wrappo. | m | 207 | | | |
| | Stools and thrust blocks | | | | | |
| L721 | Volume 0.1 - 0.2 m3 concrete grade G21 pipe nominal bore not exceeding 200 mm. | nr | 4 | | | |
| L732 | Volume 0.2 - 0.5 m3 concrete grade G21 pipe nominal bore 200 - 300 mm with strap as drawing 3/27 detail 5. | nr | 2 | | | |
| L822 | Isolated pipe support; height 1 - 1.5 m pipe nominal bore 200 - 300 mm; 1200 mm long hanger, 6 mm galvanised mild steel as drawing 3/27 detail 3. | nr | 3 | | | |
| | | | | PAGE TOTAL | | |

# *Class M: Structural metalwork*

The measurement rules for structural metalwork are relatively simple and do not generate a large number of items. Structural steelwork is a predominantly manufactured commodity. The proportion of its cost derived from processes carried out off Site in the fabrication shop is considerable. Consequently, the pricing of structural steelwork is very much concerned with estimating the cost of passing the various members through the fabrication processes required. This cost depends on the complexity of the design and on the arrangements and equipment which are available to the particular tendering company. The cost of welding on a fillet or of drilling a series of holes may differ considerably from one company to another, depending on whether and how much the particular process is mechanized or automated in each workshop. This means that the details of the shapes and sizes of members and of the connections between them have a big but varying impact on cost. It is therefore essential for drawings showing these details to be made available to companies tendering for structural steelwork in civil engineering contracts. The CESMM assumes that such drawings are supplied to tenderers and does not provide rules of measurement which are appropriate when detailed drawings are not supplied. In this context, 'detailed drawings' means dimensioned layouts indicating the sizes of members and showing details of connections and other fittings.

The second edition of the CESMM introduces few changes to the measurement of structural metalwork. The addition of items for portal frames (M 3–4 3 *) and some detailed rearrangement of the other rules are the only noticeable changes.

Much pricing for structural steelwork is carried out from a detailed materials abstract list which is produced by the tenderer from a thorough analysis of the Drawings. This list amounts to a take-off of the quantities, drawing by drawing, piece by piece, with fittings (including connections) listed against the members to which they are attached. The make-up of the itemized quantities shown on the list is used for material pricing. This list, the Drawings and the Specification are used for labour estimating in all its elements, for work in the drawing office, in the works, for fabrication and for erection.

The compiler of a bill of quantities will normally take off the quantities in a form close to that required by the tenderer as the basis of the materials abstract list. It is sensible for the results of this detailed analysis by the bill compiler to be made available to tenderers so that they do not all have to repeat the analysis. It is recommended that, on request, copies of the bill compiler's take-off list should be passed to tenderers.

The detailed materials abstract list is necessary because of the varied combination of factors which must be considered when estimating for structural steelwork. Some of the main factors are now mentioned.

(*a*) *Material* for structural steel is purchased by the tonne, but the rate per tonne varies according to the section, size (of which there are about 220), length, quality, quantity of any one size, finish and the requirements for testing and inspection. The rate per tonne for steel plates varies similarly, but is also dependent on length, width and thickness.

(*b*) *Shop and site bolts* are expensive both in material and labour and have to be estimated from an accurate forecast of the number, size and type required.

(*c*) *Drawing office costs* depend on the number of structural pieces, like and unlike, on their complexity and on the number of drawings required.

(*d*) *Fabrication costs and tonnage rates* involve labour estimating for such operations as marking off, sawing, end milling, drilling, bolting, welding, handling, assembling, fitting, straightening, preheating, testing, inspection and painting. The degree of repetition affects cost considerably.

(*e*) *Erection costs* depend on the number, size and mass of pieces, their locations and connections, as well as on site facilities and conditions. Access arrangements, availability of storage and working areas, the sequence of work and number of visits are all factors which significantly affect costs. They influence the number, type and size of cranes and other items of plant required as well as the size of the site labour force and the skills needed.

So that the compiler's take-off can be used by tenderers to assist in estimating, it is helpful to follow uniform procedures in the preparation of take-offs. The following procedure is recommended. The numbers, dimensions, masses, areas and quantities for each member should be entered on separate sheets and each entry should be described using rules such as the following.

(*a*) The numbers of the Drawings from which the quantities are taken should be stated at the beginning of the dimensions and calculations for each part of the work.
(*b*) The quantities should be listed under headings taken from the classification table in class M of the CESMM.
(*c*) Each type of member, or group of members should be described briefly and given identification marks as on the Drawings.
(*d*) Each different material component should be listed separately with the following stated: the number required, the type of section, the size of section, the unit mass or thickness, the length, the surface area (where required for measurement of surface treatments) and the mass (kg).
(*e*) The total mass of the members comprised of the components listed should be given in tonnes. The total surface area should also be given where required.
(*f*) The grade of steel in the components should be stated if it is different from the main description and the mass of the different grade of steel should be given separately.
(*g*) The sizes of steel sections stated should be as follows: actual size for angles and hollow sections, serial size for universal beams, universal columns and universal structural tees, nominal size for rolled steel joists, rolled steel channels and

rolled steel tees. Serial size is the size designation of a particular range of sections obtainable at different masses per metre of length and consequently having slightly different actual dimensions. For example, the serial sizes and masses per metre of two universal sections are 203 mm × 133 mm × 25 kg/m and 203 mm × 133 mm × 30 kg/m. As the masses are different the actual dimensions vary; they are 203·2 mm × 133·4 mm and 206·8 mm × 133·8 mm respectively. Nominal size is the actual size rounded to the nearest millimetre.

(*h*) The components of fittings should be listed and their masses shown in detail and reference given to the appropriate main member to which they are attached.

(*i*) Fittings comprise caps, bases, haunches, gussets, end plates, splice plates, cleats, brackets, stiffeners, distance pieces, separators and packings. Fittings for other trades should be identified and related to the member to which they are attached.

(*j*) Connections are fittings used to form a load-carrying joint between one member and another or to strengthen a member at such a joint.

In the second division of the Work Classification the terms trusses, purlins and cladding rails apply where these components are used in engineering structures such as conveyor gantries.

Where descriptions are called for by the additional description rules of class M it is important that separate items for different descriptions should be given in the bill. Trestles, towers, built-up columns, trusses and built-up girders can be made from sections and/or plates and may be in the form of compounded sections, lattice girders, plate girders or box-type construction. Details of these should be given to comply with rule A4.

It is trade practice to list the booms or flanges of girders and the trusses and legs of trestles first, then the internal members or web plates and last the fittings of each member or group of similar members. Camber and curvature are noted. Tapered, cranked and castellated members are taken off separately.

It is usual to secure crane rails for light cranes to their supporting beams or girders in the workshop. The crane rails are then shown with the supporting beams or girders in the materials abstract list.

The rails for heavier cranes (over 20 t capacity) usually require fixing clips and resilient pads. In these cases the work is usually done on Site by specialists and the rails are therefore listed separately.

Where materials abstract lists are supplied to tenderers they should be accompanied by a disclaimer of responsibility for their accuracy and an instruction that they should not be taken into account in the interpretation of the Contract. This is necessary so that there should be no ambiguity in the Contract in the event of error.

The other members referred to in the classification table at M 4 * * and M 7 * * are isolated, peculiar or special members itemized according to the relevant divisions and descriptions. They may be permanent or temporary. An isolated column, beam or bracing attached to a non-metal or existing metal structure is an example of a member which is part of neither a bridge nor a frame.

If blast cleaning is specified for surface treatment the standard of finish required should be stated, for example as second quality to BS 4232.* Details of painting systems should also be made clear.

Where tests are required, items are given in class A for general items at A 2 5 0 for testing of materials and A 2 6 0 for testing of the Works. Requirements for destructive and non-destructive tests, procedure tests (welding and flame cutting), qualification and testing of welders, run-off production tests and inspections should be stated clearly.

The delivery, unloading, erection and dismantling of cranes and plant can be either priced in the rates for erection of structural steel or entered as a Fixed Charge in class A at the option of the tenderer. Similarly, the cost of operating cranes and plant can either be priced in the rates for erection or entered as a Time-Related Charge in class A. Other erection costs should then be covered by the prices entered against the items for erection given in class M at M 5–7 1–2 0.

Notice that the CESMM does not require separate items to be given for fittings to structural metalwork, other than anchorages and

*text continues on page 238*

* *Surface finish of blast-cleaned steel for painting*. British Standards Institution, London, 1967, BS 4232.

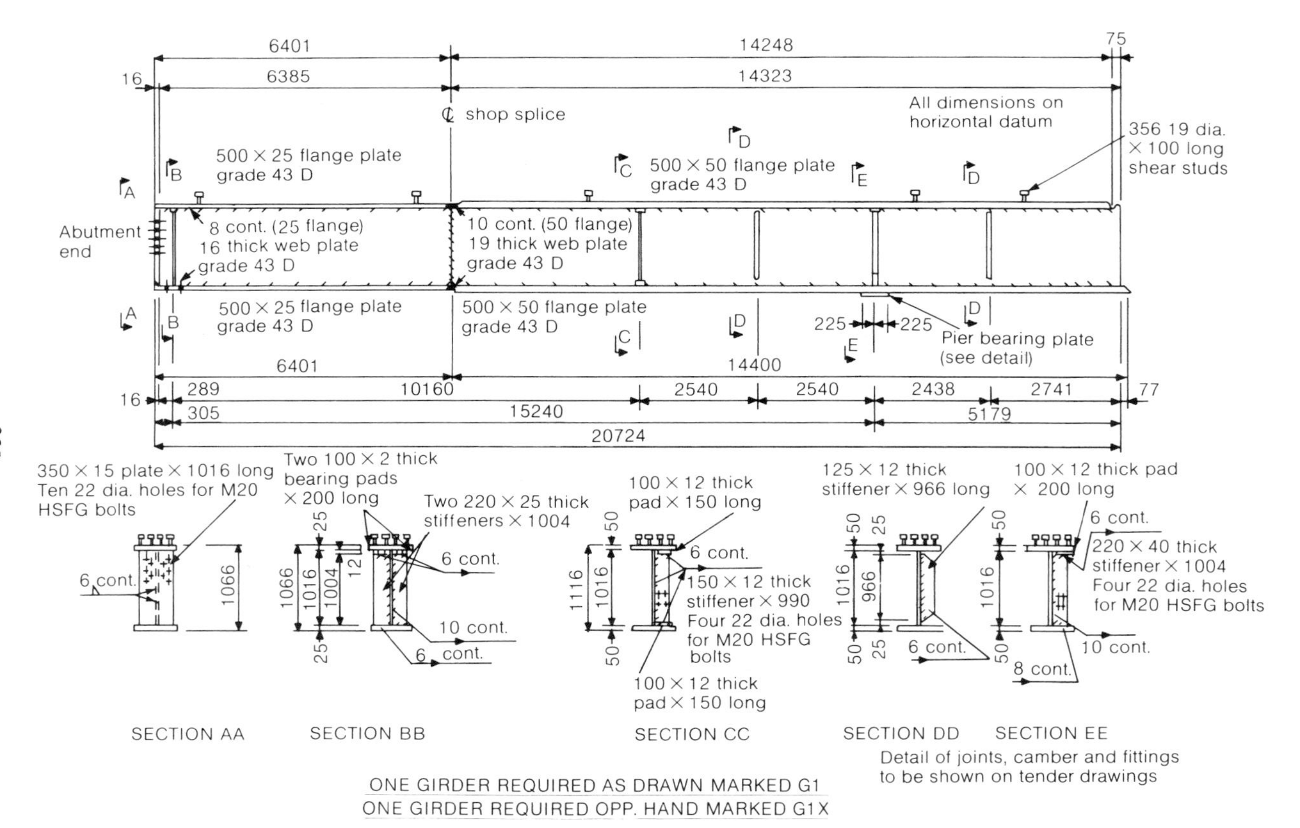


Fig. 23. Drawing of structural steel members: example A

CLIENT ______ ESTIMATE No ______

DESCRIPTION ______ DATE ______ SHEET No ______

| | No REQD. | SECTION | SIZE | LENGTH m mm | SURFACE AREA $m^2$ | MASS kg | TONNES |
|---|---|---|---|---|---|---|---|
| | | PLATE GIRDER BRIDGE DRAWING A1 | | | | | |
| | | PART ITEM M123 | | | | | |
| | | PLATE GIRDERS MK 1/G1 1/G1X | | | | | |
| TOP FLG | 1 | PLT | 500 × 50 * 43D | 14 248 | 14.25 | 2796 | |
| BTM " | 1 | " | 500 × 50 | 14 400 | 14.40 | 2826 | |
| TOP/BTM FLGS | 2 | " | 500 × 25 | 6 401 | 12.80 | 1256 | |
| WEB | 1 | " | 1016 × 19 | 14 323 | 29.10 | 2170 | |
| ↓ | 1 | " | 1016 × 16 | 6 385 | 12.97 | 815 | |
| ABUTMENT STIFFS | 2 | " | 220 × 25 | 1 004 | 0.88 | 87 | |
| ABUTMENT PADS | 2 | " | 100 × 12 | 200 | 0.08 | 4 | |
| PIER STIFF | 1 | " | 220 × 40 | 1 004 | 0.44 | 69 | |
| " PAD | 1 | " | 100 × 12 | 200 | 0.04 | 2 | |
| INTERM STIFF | 1 | " | 150 × 12 | 992 | 0.30 | 14 | |
| " PADS | 2 | " | 100 × 12 | 150 | 0.06 | 3 | |
| " STIFFS | 2 | " | 125 × 12 | 966 | 0.48 | 23 | |
| END PLT | 1 | " | 350 × 15 | 1 016 | 0.71 | 42 | |
| BEARING PLT | 1 | " | 450 × 60 | 450 | 0.41 | 95 | |
| | | | | | 86.92 | 10202 | |
| | 356 | SHEAR CONNS | 19mm DIA × 100mm LG | | | 99 | |
| | | | | IN 1 | 86.92 | 10301 | |
| | | | | IN 2 | 173.84 | 20602 | 20 602 |
| | | | * TOTAL 43D | | | 9863 | 9 863 |
| | | | SHOP BOLTS (TO THIS DRAWING) | | | NIL | |
| | | | SITE BOLTS (TO THIS DRAWING) | | | NIL | |
| | | | CARRY FORWARD | | | | |

ALL STEEL GRADE 43A UNLESS NOTED

Materials abstract list for example A shown in Fig. 23

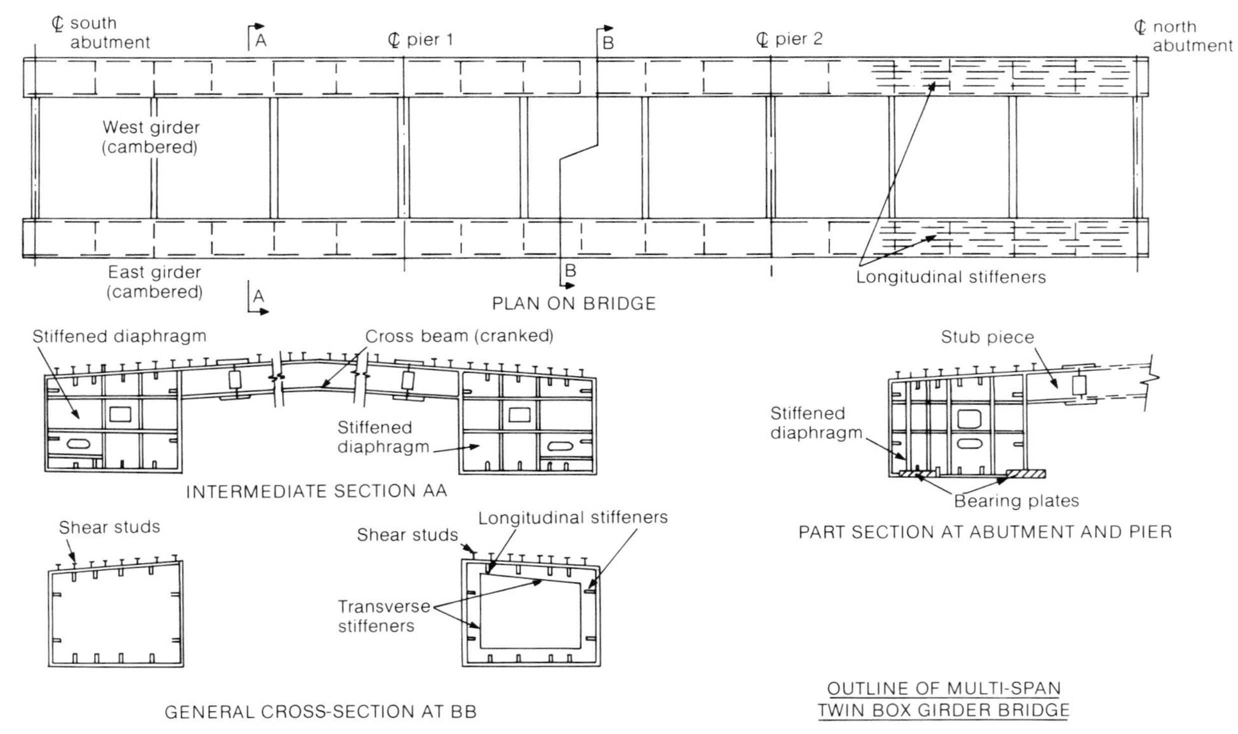


Fig. 24. Drawing of structural steel members: example B

CLIENT ESTIMATE No

DESCRIPTION DATE SHEET No

| No REQD. | SECTION | SIZE | LENGTH m mm | SURFACE AREA $m^2$ | MASS kg | TONNES |
|---|---|---|---|---|---|---|
| | TYPICAL MATERIAL TAKE-OFF SEQUENCE FOR BOX GIRDER BRIDGE | | | | | |
| | ITEM M133 | | | | | |
| | BOX GIRDER MK B DRAWING B1 | | | | | |
| | TOP FLANGE PLATES | | | | | |
| | BOTTOM FLANGE PLATES | | | | | |
| | WEB PLATES | | | | | |
| | ABUTMENT DIAPHRAGMS - INTERNAL - INCLUDING STIFFENING THERETO | | | | | |
| | PIER " - " - " " " | | | | | |
| | ITERMEDIATE " - " - " " " | | | | | |
| | LONGITUDINAL STIFFENING TO :— | | | | | |
| | TOP FLANGE | | | | | |
| | BOTTOM FLANGE | | | | | |
| | WEBS | | | | | |
| | TRANSVERSE STIFFENING TO :— | | | | | |
| | TOP FLANGE | | | | | |
| | BOTTOM FLANGE | | | | | |
| | WEBS | | | | | |
| | FITTINGS, MANHOLES AND THE LIKE | | | | | |
| | ITEM M111 | | | | | |
| | CROSS BEAMS FROM SECTIONS INCLUDING STUBS | | | | | |
| | SHEAR CONNECTORS | | | | | |
| | SPECIAL STEELS | | | | | |
| | | CARRY FORWARD | | | | |

ALL STEEL GRADE 43A UNLESS NOTED

Materials abstract list for example B shown in Fig. 24

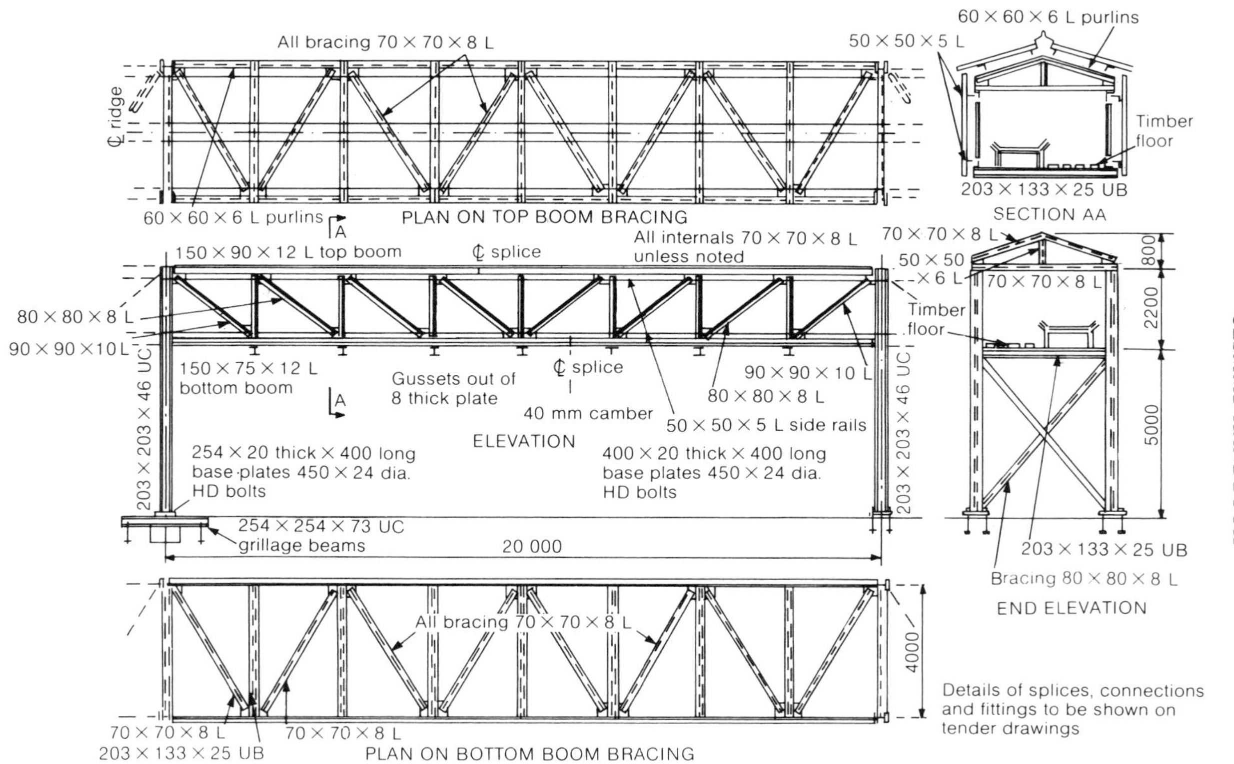


Fig. 25. Drawing of structural steel members: example C

CLIENT ______ ESTIMATE No ______

DESCRIPTION ______ DATE ______ SHEET No 1

| | No REQD. | SECTION | SIZE | LENGTH m | LENGTH mm | SURFACE AREA $m^2$ | MASS kg | TONNES |
|---|---|---|---|---|---|---|---|---|
| | | CONVEYOR GANTRY DRAWING C1 | | | | | | |
| | | ITEM M 311 | | | | | | |
| | | COLUMNS | | | | | | |
| LEG | 1 | UC | 203 × 203 × 46 | 7 | 172 | 8·51 | 330 | |
| BASE | 1 | PLT | 400 × 20 | | 400 | 0·32 | 25 | |
| CAP | 1 | " | 203 × 8 | | 203 | 0·08 | 3 | |
| | | | | IN | 1 | 8·91 | 358 | |
| | | | | IN | 2 | 17·82 | 716 | 716 |
| LEG | 1 | UC | 203 × 203 × 46 | 7 | 172 | 8·51 | 330 | |
| BASE | 1 | PLT | 254 × 20 | | 400 | 0·20 | 16 | |
| CAP | 1 | " | 203 × 8 | | 203 | 0·08 | 3 | |
| | | | | IN | 1 | 8·79 | 349 | |
| | | | | IN | 2 | 17·58 | 698 | 698 |
| | | ITEM M 321 | | | | | | |
| | | BEAMS | | | | | | |
| DECK BEAMS | 7 | UB | 203 × 133 × 25 | 4 | 164 | 26·35 | 728 | 728 |
| DECK BEAM AT COLUMNS | 1 | UB | 203 × 133 × 25 | 3 | 977 | 3·60 | 99 | |
| | 2 | PLTS | 135 × 8 | | 203 | 0·11 | 3 | |
| | | | | IN | 1 | 3·71 | 102 | |
| | | | | IN | 2 | 7·42 | 204 | 204 |
| | | | CARRY FORWARD | | | | | |

ALL STEEL GRADE 43A UNLESS NOTED

Materials abstract list for example C shown in Fig. 25

CLIENT ______ ESTIMATE No ______

DESCRIPTION ______ DATE ______ SHEET No 2

| | No REQD. | SECTION | SIZE | LENGTH m | LENGTH mm | SURFACE AREA $m^2$ | MASS kg | TONNES |
|---|---|---|---|---|---|---|---|---|
| | | CONVEYOR GANTRY (CONTD) | | | | | | |
| | | ITEM M351 | | | | | | |
| | | TRUSSES | | | | | | |
| RAFTER | 2 | ANGLES | 70 × 70 × 8 | 2 | 200 | | | |
| BTM TIE | 1 | " | " " " | 4 | 203 | 2.41 | 72 | |
| INTERM TIE | 1 | " | 50 × 50 × 6 | | 640 | 0.13 | 3 | |
| GUSSETS | 2 | PLTS | 200 × 8 | | 203 | 0.16 | 5 | |
| ↓ | 1 | " | 260 × 8 | | 360 | 0.19 | 6 | |
| | 1 | " | 200 × 8 | | 260 | 0.10 | 3 | |
| SHOE CLTS | 2 | ANGLES | 70 × 70 × 8 | | 203 | 0.11 | 3 | |
| PURLIN CLTS | 4 | " | 75 × 50 × 6 | | 150 | 0.15 | 3 | |
| | | | | IN | 1 | 3.25 | 95 | |
| | | | | IN | 9 | 29.25 | 855 | 855 |
| | | ITEM M353 | | | | | | |
| | | BUILT UP GIRDERS | | | | | | |
| TOP BOOM | 1 | ANGLE | 150 × 90 × 12 | 11 | 149 | | | |
| ↓ | 1 | " | " " " | 8 | 649 | 9.50 | 428 | |
| BTM BOOM | 1 | " | 150 × 75 × 12 | 11 | 149 | | | |
| ↓ | 1 | " | " " " | 8 | 649 | 8.91 | 400 | |
| INTERNALS | 7 | " | 70 × 70 × 8 | 1 | 900 | 3.72 | 111 | |
| | 2 | " | 80 × 80 × 8 | 3 | 000 | 1.92 | 58 | |
| | 4 | " | 70 × 70 × 8 | 3 | 000 | 3.36 | 100 | |
| ↓ | 2 | " | 90 × 90 × 10 | 2 | 900 | 2.09 | 78 | |
| GUSSETS | 1 | PLT | 240 × 8 | | 400 | 0.19 | 6 | |
| | 12 | " | 240 × 8 | | 240 | 1.38 | 43 | |
| ↓ | 1 | " | 240 × 8 | | 200 | 0.10 | 3 | |
| | 2 | UB | 203 × 203 × 46 | | 330 | | | |
| | 2 | " | " " " | | 150 | 1.14 | 44 | |
| | | ITEM M353 | CONTD SHEET 3 | | C/F | 32.31 | 1271 | |
| | | | CARRY FORWARD | | | | | |

ALL STEEL GRADE 43A UNLESS NOTED

CLIENT

ESTIMATE No

DESCRIPTION

DATE

SHEET No 3.

| | No REQD. | SECTION | SIZE | LENGTH m | LENGTH mm | SURFACE AREA m² | MASS kg | TONNES |
|---|---|---|---|---|---|---|---|---|
| | | CONVEYOR GANTRY (CONTD) | | | | | | |
| | | | ITEM M353 | | B/F | 32.31 | 1271 | |
| SPLICE PLTS | 2 | PLTS | 150 × 10 | | 350 | 0.21 | 8 | |
| | 2 | " | 125 × 10 | | 350 | 0.18 | 7 | |
| | 1 | PLT | 90 × 10 | | 350 | 0.06 | 2 | |
| | 1 | " | 75 × 10 | | 350 | 0.05 | 2 | |
| RAIL CLTS | 18 | ANGLES | 75 × 50 × 6 | | 150 | 0.68 | 15 | |
| | | | | IN | 1 | 33.49 | 1305 | |
| | | | | IN | 2 | 66.98 | 2610 | 2610 |
| | | ITEM M361 | | | | | | |
| | | BRACINGS | | | | | | |
| TRUSS TIE LEVEL | 8 | ANGLES | 70 × 70 × 8 | 4 | 580 | 10.26 | 306 | |
| | 7 | PLTS | 240 × 8 | | 400 | 1.34 | 42 | |
| | 2 | " | 200 × 8 | | 240 | 0.19 | 6 | |
| | | | | | | 11.79 | 354 | 354 |
| DECK LEVEL | 8 | ANGLES | 70 × 70 × 8 | 4 | 800 | 10.75 | 321 | |
| | 7 | PLTS | 240 × 8 | | 320 | 1.08 | 34 | |
| | 2 | " | 160 × 8 | | 248 | 0.16 | 5 | |
| | | | | | | 11.99 | 360 | 360 |
| VERTICAL | 2 | ANGLES | 80 × 80 × 8 | 6 | 000 | 3.84 | 116 | |
| | 4 | UB | 254 × 146 × 31 | | 200 | 0.84 | 25 | |
| | | | | IN | 1 | 4.68 | 141 | |
| | | | | IN | 2 | 9.36 | 282 | 282 |
| | | ITEM M361 | CONTD SHEET 4 | | C/F | 33.14 | | 496 |
| | | | | CARRY FORWARD | | | | |

ALL STEEL GRADE 43A UNLESS NOTED

CLIENT

ESTIMATE No

DESCRIPTION

DATE

SHEET No 4

| | No REQD. | SECTION | SIZE | LENGTH m | LENGTH mm | SURFACE AREA $m^2$ | MASS kg | TONNES |
|---|---|---|---|---|---|---|---|---|
| | | CONVEYOR GANTRY (CONTD) | | | | | | |
| | | ITEM M361 | | | B/F | 33.14 | | 996 |
| | | PURLINS | | | | | | |
| | 16 | ANGLES | 60 × 60 × 6 | 5 | 000 | 19.20 | 434 | 434 |
| | | CLADDING RAILS | | | | | | |
| | 16 | ANGLES | 50 × 50 × 5 | 5 | 000 | 16.00 | 302 | 302 |
| | | | TOTAL | | | 68.34 | | 1732 |
| | | ITEM M 370 | | | | | | |
| | | GRILLAGE BEAMS. | | | | | | |
| | 2 | UC | 254 × 254 × 73 | 2 | 500 | 7.43 | 365 | 365 |
| | | ITEM M 380 | | | | | | |
| | | HOLDING DOWN BOLT ASSEMBLIES. | | | | | | |
| | 16 | H.D BOLTS | M24 SQ/SQ/OX | | 450 | | 35 | |
| | 16 | PLTS. | 150 × 10 | | 150 | | 28 | |
| | | | | | | | 63 | 063 |
| | | | CARRY FORWARD | | | | | |

ALL STEEL GRADE 43A UNLESS NOTED

CLIENT

ESTIMATE No

DESCRIPTION

DATE

SHEET No 5

| No REQD. | SECTION | SIZE | LENGTH m | LENGTH mm | SURFACE AREA $m^2$ | MASS kg | TONNES | |
|---|---|---|---|---|---|---|---|---|
| | CONVEYOR GANTRY | | | | | | | |
| | SUMMARY | | | | | | | |
| | | ITEM M 311 | | | 35.40 | | 1 | 414 |
| | | ITEM M321 | | | 33.77 | | | 932 |
| | | ITEM M351 | | | 29.25 | | | 855 |
| | | ITEM M353 | | | 66.98 | | 2 | 610 |
| | | ITEM M361 | | | 68.34 | | 1 | 732 |
| | | ITEM M370 | | | 7.43 | | | 365 |
| | | TOTAL | | | 241.17 | | 7 | 908 |
| | | ITEM M 380 | | | | | 4 | No |
| | SHOP | BOLTS M20 × 50 XOX BLACK | | | | | 116 | No |
| | | ITEM M 632 | | | | | | |
| | SITE | BOLTS M20 × 50 XOX BLACK | | | | | 150 | No |
| | | ITEM M 662 | | | | | | |
| | SITE | BOLTS M20 × 60 XOX HSFG | | | | | 84 | No |
| | | CARRY FORWARD | | | | | | |

ALL STEEL GRADE 43A UNLESS NOTED

bolts. In accordance with rule M3, the mass for the main items includes the mass of fittings, and the prices entered against the main items should cover the work of providing the fittings. It is important that the Drawings and Specification define the fittings required as the tenderer will allow in his prices only for work which is so defined.

Figures 23–25 show three examples of structural steel members. Materials abstract lists in the recommended form for taking off are shown for each example on pages 229, 231 and 233–237 and the resulting bill items for the example in Fig. 25 are shown in the example bill pages. The quantities against these items are entered to the nearest tenth of a tonne, in accordance with paragraph 5.18. The rules of the CESMM do not prohibit quantities being given with greater precision and these items may be some of the few which justify a second decimal place. However, it should be borne in mind that, even with prices in the region of £800/t, the maximum error in extension of one item as a result of not using the second decimal place is £40, the average error is £20 and the errors which do occur will be mutually compensating unless there are only a very few items in this class.

*Schedule of changes in CESMM2*

1. Classification of main and subsidiary members is moved to the first division.
2. Items for portal frames are added.
3. Surface treatment items are confined to off-Site work.

| Number | Item description | Unit | Quantity | Rate | Amount £ | p |
|---|---|---|---|---|---|---|
| | STRUCTURAL METALWORK. | | | | | |
| | Conveyor gantry example C, steel grade 43A. | | | | | |
| | Fabrication of members for frames straight on plan. | | | | | |
| M311 | Columns. | t | 1.4 | | | |
| M321 | Beams. | t | 0.9 | | | |
| M351 | Roof trusses comprising single 70 x 70 x 8 angle rafters and 50 x 50 x 6 internal and bottom ties. | t | 0.9 | | | |
| M353 | Built-up side girders cambered comprising two single 150 x 90 x 12 angles top boom, two single 150 x 75 x 12 angles bottom boom with verticals 70 x 70 x 8 angles, diagonals 70 x 70 x 8, 80 x 80 x 8 and 90 x 90 x 10 angles. | t | 2.6 | | | |
| M361 | Bracings, purlins and cladding rails. | t | 1.7 | | | |
| M370 | Grillages. | t | 0.4 | | | |
| M380 | Anchorages and holding bolt assemblies comprising 4 nr 450 x 24 bolts with plates 150 x 150 x 10. | nr | 4 | | | |
| | Erection of conveyor gantry. | | | | | |
| M620 | Frame members. | t | 7.9 | | | |
| M632 | Site bolts black diameter 16 - 20 mm. | nr | 150 | | | |
| M662 | HSFG load indicating bolts diameter 16 - 20 mm with washers. | nr | 84 | | | |
| | | | PAGE TOTAL | | | |

| Number | Item description | Unit | Quantity | Rate | Amount £ | p |
|---|---|---|---|---|---|---|
| | STRUCTURAL METALWORK | | | | | |
| | Conveyor gantry example C, steel grade 43A. | | | | | |
| | Off Site surface treatment. | | | | | |
| M810 | Blast cleaning to BS 4232 second quality. | m2 | 241 | | | |
| M870 | Painting one coat zinc epoxy primer. | m2 | 241 | | | |
| | | | | PAGE TOTAL | | |

# *Class N: Miscellaneous metalwork*

Class N includes all metal components which are not specifically included in another class. It is the class into which fall the purpose-made odd bits of metalwork, some of which are required in most civil engineering contracts but which have no generic name. This is why the first division of the classification table does not give names to its descriptive features. The different things listed in the second division do not fall into groups which could usefully be identified or named.

The itemization of miscellaneous metalwork is not elaborate; separate items are not given for erection or fixing, or for fixings themselves. The class reverts to the traditional shopping list approach to measurement because it remains appropriate to this type of work. Only minor changes to class N are made in CESMM2.

The note at the foot of page 61 of CESMM2 is a good example of the principle adopted in the CESMM with regard to the role of drawings. Rule A1 requires that a comprehensive description of each miscellaneous metalwork assembly shall be given. In the case of something composed of many pieces, like a staircase, this could be a very lengthy and a detailed description. The note gives the alternative of referring to a mark number which identifies an assembly the details of which are shown on the Drawings or given in the Specification. This alternative should be well used as there is no point in giving a lengthy description which in the event is only used by the tenderer or supplier to identify work the details of which he derives from the Drawings. This principle, which is also adopted in, for instance, class H for precast concrete, is quite different from the approach adopted in the building method of measurement where it is assumed that all information relevant to prices for work is conveyed in the bill item descriptions.

The classification table for class N has little impact on the measurement of miscellaneous metalwork. Its main function is to provide classification numbers for some of the more commonly occurring work in the class and to provide special measurement units for such items when a departure from the normal mass measurement is warranted. Non-standard components will normally be measured by the tonne and will be identified by reference to a drawing. Rule C1 means that miscellaneous metalwork items normally include fixing. Where this is not the case, item descriptions must say so as required by paragraph 3.3.

Metal ladders are measured by their length. Where the stringers at the top of a ladder are extended and returned to form a handrail the length of the ladder is taken to include the full length of the stringer including its return.

The second edition of CESMM introduces new items for cladding (N 2 1 0). This is intended to be used for relatively simple cladding encountered in civil engineering contracts. Complex cladding encountered in building work should be measured in more detail.

*Schedule of changes in CESMM2*

1. Rectangular frames are renamed miscellaneous framing.
2. Items for cladding are added.
3. Walings are measured by length.
4. Tank covers are not measured separately.

| Number | Item description | Unit | Quantity | Rate | Amount | |
|---|---|---|---|---|---|---|
| | | | | | £ | p |
| | MISCELLANEOUS METALWORK. | | | | | |
| N110.1 | Stairways and landings; staircase S3 drawing 136/27. | t | 3.7 | | | |
| N110.2 | Stairways and landings; staircase S4 drawing 136/28. | t | 4.5 | | | |
| N130.1 | Galvanised mild steel ladders; stringers 65 x 13 mm, 400 mm apart, rungs 20 mm diameter at 300 mm centres; stringers extended and returned 1000 mm to form handrail. | m | 14 | | | |
| N130.2 | Galvanised mild steel ladders; stringers 65 x 13 mm, 400 mm apart, rungs 20 mm diameter at 300 mm centres; stringers extended and returned 1000 mm to form handrail; with safety cage of 3 nr 65 x 13 mm verticals and 65 x 13 mm hoops 750 mm diameter at 700 mm centres cage commencing 2500 mm above commencing level. | m | 24 | | | |
| N161.1 | Miscellaneous framing; galvanised mild steel angle section 64 x 64 x 9 mm. | m | 247 | | | |
| N161.2 | Miscellaneous framing; galvanised mild steel angle section 38 x 38 x 6 mm. | m | 53 | | | |
| N170 | Plate flooring; galvanised mild steel chequer plating 10 mm thick as detailed on drawing 136/42. | m2 | 86 | | | |
| N180 | Open grid flooring; galvanised mild steel to BS 4592 as detailed on drawing 137/49. | m2 | 47 | | | |
| N230 | Duct covers; galvanised mild steel 10 mm thick, width 300 mm as detailed on Drawing 137/50. | m2 | 27 | | | |
| N286 | Covered tanks volume 100 - 300 m3; galvanised mild steel to BS 417 part 1 reference T40. | nr | 3 | | | |
| | | | | PAGE TOTAL | | |

| Number | Item description | Unit | Quantity | Rate | Amount £ | p |
|---|---|---|---|---|---|---|
| | MISCELLANEOUS METALWORK. | | | | | |
| N999.1 | Galvanised mild steel adjustable V-notch weir plate to precise levels as detailed on drawing 136/7 detail A. | m | 137 | | | |
| N999.2 | Cast iron light duty inspection cover to BS 497 reference C6 - 24/24. | nr | 2 | | | |
| N999.3 | Supply cast iron step irons to BS 1247 type a2. | nr | 12 | | | |
| N999.4 | Supply 2200 x 2200 mm galvanised mild steel forebay screen as detailed on drawing 136/98. | t | 1.4 | | | |
| | | | PAGE TOTAL | | | |

# *Class O: Timber*

The effect of the includes and excludes list of class O is that the rules for measurement given are intended to apply only to timber components used for Permanent Works in civil engineering contracts. Decking and fendering in harbour work are examples. The rules are not intended to apply to small scale carpentry and joinery such as windows, doors and skirtings in buildings. Where any such work is included in a civil engineering contract it should be billed by simple identification related to the Drawings and Specification and the components should be measured by number or by length. If non-standard measurement conventions have to be used they should be defined in the Preamble to the bill and if, where warranted by the amount of such work, a different standard method of measurement is used this also should be referred to in the Preamble.

The measurement rules for timber are straightforward. Decking is measured in square metres and all other timber components are measured by their length in metres. Fittings and fastenings are enumerated. Little change to the measurement rules for timber were made when the second edition of the CESMM was introduced. 'For maritime use' was dropped as a classification in division one in favour of stating the grade of timber required (rule A1).

Item descriptions must describe the timber grade and species and give nominal gross sizes in each case. Nominal gross sizes are the sizes of the sawn pieces of timber from which the required components are produced. They exceed finished sizes by the thicknesses of material removed during planing or other finishing processes.

Rule M2 establishes that the area of timber decking measured includes the area of holes and openings which are each less than $0{\cdot}5\ m^2$ in area. This means that if timber decking is blocked to provide narrow openings between strips of decking timber normally it will be measured without deducting the area of the openings. Gaps 10 mm wide by as much as 49 m long are less than 0·5 $m^2$ in area. This arrangement occurs in some marine work where decking is intended to be self-draining.

In CESMM2, materials, types and sizes are to be stated in item descriptions for timber fittings and fastenings (rule A4).

*Schedule of changes in CESMM2*

1. Marine use of timber is not separately categorized.
2. Details of fittings and fastenings are to be stated.

| Number | Item description | Unit | Quantity | Rate | Amount £ | p |
|---|---|---|---|---|---|---|
| | TIMBER. | | | | | |
| | Hardwood components cross-sectional area 0.01 - 0.02 m2; wrought finish. | | | | | |
| O122 | 150 x 75 mm Greenheart length 1.5 - 3 m. | m | 184 | | | |
| O123 | 150 x 100 mm Greenheart length 3 - 5 m; pier decking runners. | m | 204 | | | |
| | Hardwood components cross-sectional area 0.04 - 0.1 m2; wrought finish. | | | | | |
| O143 | 300 x 300 mm Greenheart length 3 - 5 m; pier braces between piles. | m | 163 | | | |
| | Hardwood components cross-sectional area 0.1 - 0.2 m2; wrought finish. | | | | | |
| O152 | 350 x 350 mm Greenheart length 1.5 - 3 m. | m | 287 | | | |
| O153 | 400 x 400 mm Greenheart length 3 - 5 m; pier decking bearers. | m | 364 | | | |
| | Softwood components cross-sectional area not exceeding 0.01 m2; wrought finish. | | | | | |
| O211 | 75 x 75 mm Douglas fir tanalized length not exceeding 1.5 m. | m | 196 | | | |
| O212 | 100 x 100 mm Douglas fir tanalized length 1.5 - 3m. | m | 47 | | | |
| O213 | 100 x 50 mm Douglas fir tanalized with rounded eges length 3 - 5 m; hand rail. | m | 280 | | | |
| | | | PAGE TOTAL | | | |

DOCK 3

| Number | Item description | Unit | Quantity | Rate | Amount £ | p |
|---|---|---|---|---|---|---|
| | TIMBER. | | | | | |
| | Softwood components cross-sectional area 0.04 - 0.1 m2; wrought finish. | | | | | |
| 0242 | 150 x 300 mm Douglas fir length 1.5 - 3 m. | m | 48 | | | |
| 0243 | 225 x 300 mm Douglas fir length 3 - 5 m; pier rubbing piece. | m | 440 | | | |
| | Hardwood decking thickness 50 - 75 mm; wrought finish. | | | | | |
| 0330.1 | 150 x 70 mm Greenheart. | m2 | 786 | | | |
| 0330.2 | 250 x 70 mm Greenheart. | m2 | 260 | | | |
| | Fittings and fastenings. | | | | | |
| | Galvanised mild steel. | | | | | |
| 0510 | Straps girth: 457 mm width 50 mm thickness 5 mm as drawing D3/27. | nr | 10 | | | |
| 0520 | Spikes: length 75 mm. | nr | 50 | | | |
| 0540.1 | Bolts: length 75 mm, diameter 5 mm. | nr | 50 | | | |
| 0540.2 | Bolts; stainless steel length 100 mm diameter 6 mm. | nr | 40 | | | |
| 0550 | Plates; stainless steel 100 x 100 mm thickness 6 mm. | nr | 10 | | | |
| | | | PAGE TOTAL | | | |

# *Class P: Piles*

The rules for measurement of piling work demonstrate most of the principles and developments embodied in the CESMM. The bill items and measurements which they generate are intended to provide a price breakdown for piling work which reflects realistically how its costs build up. The measurement of piles shows clearly the usefulness of Method-Related Charges. Some of the costs of piling operations which are suited to pricing as Fixed Charges are those of transporting plant, kentledge and equipment to the Site, and building stagings. Time-related costs of operating piling plant can be represented by Time-Related Charges. Method-Related Charges can also be used to allow for the additional costs of working in water when this occurs.

In class P items are generated which are intended to produce a set of prices for each piling operation which will lead to realistic payment in the event of variations. The piling classes were reviewed in detail for CESMM2 and a number of significant changes made.

The uncertainty which surrounds piling operations often results in Engineers' delegating some of their design responsibility to the Contractor through the medium of contractor design. The CESMM provides for measurement of contractor-designed construction in paragraph 5.4. In the case of fully designed piling work, the Contractor is still required to take on some of the risks involved as his expertise places him in the best position to evaluate the likely extent of many uncertain circumstances. The CESMM reflects this by not requiring separate items for some components of the work. An example is 'obstructions' which are separately itemized for bored piling only. Driving through obstructions is either possible or it is not and, where it is possible, it is so difficult to measure realistically that the risk is better transferred to the Contractor. Other examples of Contractor's risk items which are not separately itemized are redriving risen piles and the loss of concrete into voids in the ground when constructing cast in place piles.

A set of bill items is used to measure each group of piles. A group of piles comprises all the piles which are located together to support a single structure from the same Commencing Surface and which are all of the same type, of the same material and have the same cross-sectional characteristics. All the piles supporting a bridge, for example, would be grouped according to their type, material and size. If there were two differing Commencing Surfaces, such as the river bed surface and an embankment surface, piles driven from them would be grouped separately. Thus on a simple job all the piles might be regarded as one group, but there would be several groups on a job needing several different pile sizes and types. The items given for each group are generated by the third division of the classification table, read in conjunction with the appropriate rules. For each group of cast in place piles the depth stated in the item description for the total depth bored or driven (P 1–2 * 3) is the single depth of the deepest pile in that group (rule D1) (Fig. 26). This convention recognizes that the cost of boring or driving is, to a large extent, dependent on the type of piling plant used. This is in turn dependent on the greatest depth to be bored or driven. Where measurements are taken from the designed Commencing Surface, even though the Contractor at his discretion may adopt another Commencing Surface in carrying out the work, the original Commencing Surface will still be adopted for the purpose of measurement (rule M1).

The item for the number of piles in the group covers all the costs which are proportional to the number of piles to be constructed. These might include the plant and labour costs associated with moving the rig from one pile position to the next, setting up at each position and getting ready to drive or bore and also the cost of driving heads and pile shoes. The item for the length of piles is measured according to the total length of all the piles in the group. It covers the main material cost of the piles. In the items for preformed piles (P 3–7 * *) the length item is the length ordered or supplied (rules D3 and D4). This is so that the bill rates can model the fact that the cost per unit of length of most preformed piles varies

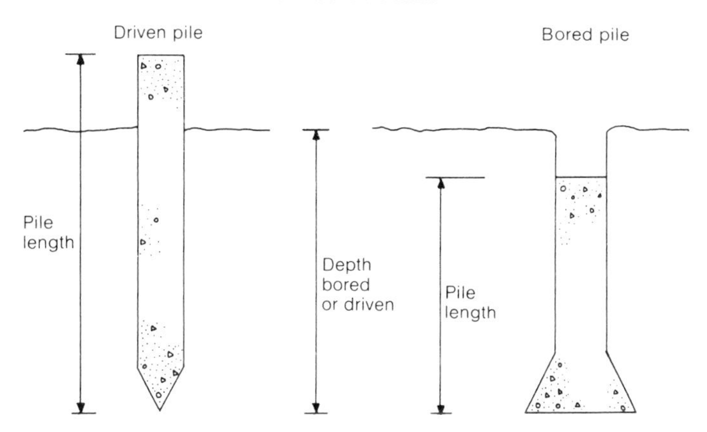


Fig. 26. Measurement of piles. At least two bill items are given for each group of preformed concrete, timber or isolated steel piles: number of piles of a stated length and depth driven. At least three bill items are given for each group of cast in place piles: number of piles, concreted length and depth bored or driven stating the depth of the deepest pile in the group. These items are generated by the third division of classification of class P and associated rules.

according to the maximum length of the pile. Also the material cost of preformed concrete piles includes reinforcement and formwork. Rule A1 requires reinforcement in such piles to be identified as it forms part of the material of the pile.

It is the Engineer's responsibility to specify the length of piles to be supplied for driving. As the price of a supplied length is governed by the length purchased or cast, it is reasonable for the precise length of a pile to be stated and priced accordingly. On one Site, only one or two lengths may be required and it is simple to state the individual lengths in the item descriptions and to give the number of lengths required. The manufactured lengths, the cost of joints to make up specified lengths and haulage govern the total supply cost. The cost of handling and pitching is a fixed cost which can be included with the enumerated item. Once the units have been pitched the driving rate of penetration is fairly constant.

Interlocking sheet piles are covered by items P8**. The same principles apply to these piles as to groups of isolated piles, except that the number of piles driven is not measured. Instead the items for driving and materials are measured by area. The areas are calculated by multiplying the depth of pile or the driven depth by the undeveloped length of the pile wall formed (see rule M7). The mean length is used where the top and bottom lengths differ because the ends of the pile walls are not vertical. In the items for the areas of piles (P 8 * 3–5) the lengths of the piles are stated with respect to classified ranges. These ranges identify the differing transport costs associated with piles of differing lengths.

The second division of classification in class P uses ranges of various cross-sectional characteristics of each type of pile. These are overridden in item descriptions by rules A4, A6, A9 and A10 which require particular dimensions to be stated for each group of piles.

Rules M1 and A3 use the term Commencing Surface to provide the rule for establishing the level at which boring for or driving piles is expected to begin. Where the Commencing Surface is not the Original Surface, the Commencing Surface must be defined in the appropriate item descriptions, and there must be an item somewhere in the bill for excavating or for filling to this surface from the Original Surface or for constructing a temporary piling platform. This is the effect on piling of paragraph 5.21.

CESMM2 uses a new definition of the lengths of cast in place piles (rule M3). It does not include any surplus length concreted at the Contractor's discretion.

The lengths stated in item descriptions for the number of preformed concrete, timber and steel piles are the lengths which are expressly required. This means that the Contractor relies on the Engineer's previous judgement when ordering lengths of preformed piles. Any alterations from these lengths will be valued by measuring surplus lengths cut off or extensions added as provided for in class Q.

Class P illustrates how the traditional mechanism of extra over items is obviated by the rules in the CESMM. The two items for each group of preformed piles—number and depth driven—both contribute to the total price for constructing the group of piles. They may be supplemented by Method-Related Charges. It might be argued that the number item is extra over the item for driven depth or vice versa, or that they are both extra over the Method-Related Charges or vice versa. It follows that to identify any one as extra over does nothing to alter the significance or coverage of the item.

Using the CESMM, a simple guide for pricing prevails in place of the subtleties of extra overs. The cost which should be allowed for in the rate inserted against an item is the cost which is proportional to the quantity which is set against the item. This guide applies whether the item is one which might formerly have been thought of as extra over or not.

Rule M7 states, among other things, that the two area measurements for interlocking steel piles include the area occupied by special piles such as tapers and corner piles. It follows that the prices per unit of length of these special piles (P 8 * 1) should only allow for any cost of labour and materials for these piles which is additional to the cost of the same area of ordinary piles. Rule M7 effectively makes the items for special piles extra over rates.

CESMM2 requires inclination ratios to be stated for raked piles (rule A2). Common ratios are 1:4, 1:6 and 1:10 which correspond to angles of 14°, 9·5° and 5·8° respectively.

The cost of disposal of surplus excavated materials is deemed to be included in rates for piling work (rule C1). This is because arisings from piling operations are not taken into account in soil balance calculations and the responsibility for disposal is usually divided between the main Contractor and the piling sub-contractor. Where there are specific requirements concerning disposal, such as whether it is to be on or off the Site, this should be given as additional description in accordance with paragraph 5.10.

*Schedule of changes in CESMM2*

1. Third division itemization is revised with simplified indication of depths.
2. Details of treatments and coatings and of pile heads and shoes are to be given in item descriptions.
3. The supported structure and the Commencing Surface are to be identified in item descriptions for piles.

PUMPING STATION

| Number | Item description | Unit | Quantity | Rate | Amount £ | p |
|---|---|---|---|---|---|---|
| | Piling to Pumping Station. | | | | | |
| | Commencing Surface 33.00 a.o.d | | | | | |
| | Bored cast in place piles concrete grade C25 as Specification clause 713.3. | | | | | |
| | Diameter 900 mm. | | | | | |
| P151 | Number of piles. | nr | 97 | | | |
| P152 | Concreted length. | m | 1195 | | | |
| P153 | Depth bored maximum depth 18 m. | m | 1305 | | | |
| | Diameter 1200 mm raked at inclination ratio 1:6. | | | | | |
| P161 | Number of piles. | nr | 42 | | | |
| P162 | Concreted length. | m | 932 | | | |
| P163 | Depth bored maximum depth 35 m | m | 992 | | | |
| | | | PAGE TOTAL | | | |

BRIDGE

| Number | Item description | Unit | Quantity | Rate | Amount £ | p |
|---|---|---|---|---|---|---|
| | Piling to Bridge. | | | | | |
| | Commenced Surface to be Original Surface. | | | | | |
| | Preformed piles concrete grade C25 as Specification clause 713.4 reinforcement as detail 4 drawing 137/65 circular diameter 300 mm. | | | | | |
| P331.1 | Number of piles length 8.5 m; mild steel driving heads and shoes. | nr | 10 | | | |
| P331.2 | Number of piles length 12.5 m; mild steel driving heads and shoes. | nr | 17 | | | |
| P332 | Depth driven. | m | 284 | | | |
| | Preformed piles concrete grade C25 as Specification clause 713.4 reinforcement as detail 5 on drawing 137/65 nominal circular diameter 450 mm. | | | | | |
| P351.1 | Number of piles length 12.5 m. | nr | 128 | | | |
| P351.2 | Number of piles length 17.2 m. | nr | 40 | | | |
| P352 | Depth driven. | m | 2120 | | | |
| | | | PAGE TOTAL | | | |

| Number | Item description | Unit | Quantity | Rate | Amount £ | p |
|---|---|---|---|---|---|---|
| | Piling to Pier. | | | | | |
| | Timber piles cross-sectional area: 0.15 - 0.25 m2; 400 x 400 mm Greenheart. | | | | | |
| | Commencing Surface to be Original Surface of bed of River Corve. | | | | | |
| P651 | Number of piles length 7.5 m; galvanised mild steel driving head and shoe as detail 5 drawing 140/7. | nr | 104 | | | |
| P652 | Depth driven. | m | 630 | | | |
| | | | | PAGE TOTAL | | |

STORAGE TANK

| Number | Item description | Unit | Quantity | Rate | Amount £ | p |
|---|---|---|---|---|---|---|
| | PILING TO STORAGE TANK. | | | | | |
| | Interlocking grade 43A steel piles Lincolnshire Steel type 2N section modulus 1150 cm3/m. | | | | | |
| | Commencing Surface 300 mm below Original Surface. | | | | | |
| P831.1 | Length of corner piles. | m | 44 | | | |
| P831.2 | Length of taper piles. | m | 27 | | | |
| P832 | Driven area. | m2 | 6847 | | | |
| P833 | Area of piles of length: not exceeding 14 m; treated with two coats bitumen paint. | m2 | 3764 | | | |
| P834 | Area of piles of length: 14 - 24 m; treated with two coats bitumen paint. | m2 | 4367 | | | |
| | | | | PAGE TOTAL | | |

# *Class Q: Piling ancillaries*

Class Q includes all work incidental to piling operations—everything other than the piles themselves. The class is unusual in that a general measurement rule (rule M1) establishes that all work in this class (except backfilling empty bores for cast in place concrete piles) is to be measured only where it is expressly required. This means that all the rest of the work will only be paid for separately if it is expressly required, not if the Contractor does any of it of his own volition. Backfilling empty bores for cast in place piles is the exception because this work is obligatory under the statutory regulations governing safety. A new general rule provides that all the prices in class Q should include for disposal of surplus materials unless otherwise stated. This is a change between the first and second editions of the CESMM. Surplus materials include cut-off lengths of piles and casings.

Work ancillary to piling is classified by pile type and size using the same terms as those used in class P to identify the piles themselves. The rules in class Q do not require precise cross-sectional dimensions to be given for the ancillaries to piles. Instead each type of pile is classified using an appropriate range of dimensions.

Permanent casings are measured in accordance with rule M2 and distinguished according to whether they exceed 13 m in length or not. This is the length which governs whether they can be transported complete or need to be divided into sections for transport.

Items for cutting off surplus lengths (Q 1–6 7 *) include those lengths of permanent casings which are surplus to requirements. Temporary casings are not measured unless the casings are expressly required and are also expressly required to be removed. The measurement of reinforcement in cast in place concrete piles is more detailed in CESMM2 than in the first edition but is still in less detail than that required for ordinary reinforcement in concrete (class G). Reinforcement in any other type of pile is regarded as a constituent material and should be described within the main pile item description. This applies, amongst other things, to filling hollow piles with concrete.

Where the base of a cast in place pile is to be enlarged, rule A1 requires that the resulting diameter should be stated. Note that the length of piles measured in the items in class P includes the length of enlarged bases (rule M3 of class P) so that the cost to be allowed for in the rate for the enlarged bases themselves is only the additional cost of the plant and labour forming the enlargement and of the volume of concrete which is either outside the cross-sectional area of the shaft or below the toe level of the shaft which is expressly required.

The arrangements for measuring pile extensions are intended to relate logically to the costs involved. An item is given for the number of pile extensions (Q 3–6 4 *). This should cover the cost of preparing the piles to receive the extensions and of making the joints. A maximum of two other items (Q 3–6 5–6 *) is given for the total length of pile extensions subdivided into those which do not exceed 3 m long and those which exceed 3 m long. These items cover the cost of the material in the pile extensions themselves. No item is given for driving extended piles as this length of driving is included in the measurement of driving against the items in class P. Where the costs of moving a piling rig to the site of a pile extension and of setting the pile in motion again are significant, they should be allowed for in the rate against the item for the number of pile extensions.

The items for preparing heads of piles in this class (Q 1–6 8 *) cover the cost of the work required under the particular Contract to prepare the head of a pile to receive subsequent permanent work. This should include breaking down of concrete if it is required. Clearly, these items do not also cover preparing the end of a pile to receive an extension as this is not preparing the head of a pile. The cost of this work should be covered in the item for the number of pile extensions.

In CESMM2, obstructions are measured in hours at Q 7 0 0. This quantity is the time taken by any piling team to break out rock or artificial hard material encountered above the founding stratum when constructing bored piles (rule M11). Bearing in mind that this item is also only measured when the work is expressly required, it is clear that the Engineer or his staff will be involved in decisions affecting the time taken to deal with obstructions. If any rock or artificial hard material is encountered when constructing bored piles which does not delay the boring work, no obstruction has taken place. The reference to the founding stratum in rule M11 is included so that normal boring into the founding stratum in order to secure the toe is not regarded as dealing with an obstruction.

The CESMM does not provide rules for the measurement of extracting piles. If such work is included in a contract it is appropriate to give items for it as non-standard work in class Q. Suitable item codes are Q 3–6 9 *.

*Schedule of changes in CESMM2*

1. Items for disposal of surplus materials, boring through rock, placing concrete by trémie pipe and standing-by items do not appear in the CESMM second edition.
2. An item is included for cutting off surplus lengths of cast in place concrete piles where expressly required.
3. Reinforcement in cast in place concrete piles is measured in more detail.
4. Pile extensions for preformed piles are categorized by length.
5. Items for pile tests are expanded and revised.
6. The rules for measurement of time spent in dealing with obstructions in bored piling are changed.

PUMPING STATION

| Number | Item description | Unit | Quantity | Rate | Amount £ | p |
|---|---|---|---|---|---|---|
| | PILING ANCILLARIES. | | | | | |
| | Cast in place concrete piles. | | | | | |
| Q125 | Backfilling empty bore with selected excavated material diameter 900 mm. | m | 20 | | | |
| Q135 | Permanent casings each length not exceeding 13 m, diameter 900 mm; 6 mm thick mild steel, bitumen painted. | m | 40 | | | |
| Q155 | Enlarged bases; 2500 mm diameter to 900 mm diameter pile. | nr | 87 | | | |
| Q156 | Enlarged bases; 3000 mm diameter to 1200 mm diameter pile. | nr | 42 | | | |
| Q185 | Preparing heads; 900 mm diameter. | nr | 87 | | | |
| Q186 | Preparing heads; 1200 mm diameter. | nr | 42 | | | |
| | Mild steel reinforcement to BS 4449. | | | | | |
| Q211 | Straight bars, nominal size not exceeding 25 mm. | t | 46.3 | | | |
| Q212 | Straight bars, nominal size: exceeding 25 mm. | t | 27.3 | | | |
| Q700 | Obstructions. | h | 150 | | | |
| | | | PAGE TOTAL | | | |

PIER

| Number | Item description | Unit | Quantity | Rate | Amount £ | p |
|---|---|---|---|---|---|---|
| | PILING ANCILLARIES. | | | | | |
| | Timber piles cross-sectional areas 0.15 - 0.25 m2. | | | | | |
| Q415 | Pre-boring. | m | 200 | | | |
| Q445 | Number of pile extensions. | nr | 50 | | | |
| Q455 | Length of pile extensions each length not exceeding 3 m. | m | 75 | | | |
| Q475 | Cutting off surplus lengths. | nr | 25 | | | |
| Q700 | Obstructions. | h | 10 | | | |
| Q811 | Pile tests, maintained loading with various reactions test load: 75 t; to working pile. | nr | 5 | | | |
| | | | PAGE TOTAL | | | |

| Number | Item description | Unit | Quantity | Rate | Amount £ | p |
|---|---|---|---|---|---|---|
| | PILING ANCILLARIES. | | | | | |
| | Interlocking steel piles section modulus 1150 cm3/m. | | | | | |
| Q643 | Number of pile extensions. | nr | 150 | | | |
| Q653 | Length of pile extensions each length: not exceeding 3m. | m | 300 | | | |
| Q673 | Cutting off surplus lengths. | nr | 150 | | | |
| Q700 | Obstructions. | h | 20 | | | |
| | | | PAGE TOTAL | | | |

# *Class R: Roads and pavings*

The provisions of class R apply equally to airport runways as to ordinary roads. A separate subclassification is given only for what are called light duty pavements (R 7 * *). This is not significantly different from the classification for heavy duty work but it will be found to be generally more appropriate for work such as footpaths, cycle tracks and what might be called domestic road-works. Little change was made to class R in revising the CESMM for the second edition.

The measurement rules for roads and pavings are basically simple with none of the intricacies necessary for the measurement of more geometrically or technically complex work. The various courses of road making materials are measured by area with kerbs and other miscellanea measured by length or enumerated. Each course of material is described and its depth or thickness stated.

The third division of the classification table in class R gives ranges of depth dimensions, but these are overridden by the requirement of rule A1 that the actual depth should be stated in item descriptions for courses of road and pavement materials and the spread rate stated for applied surface finishes.

The details of this class draw heavily on the provisions of the *Specification for road and bridge works* published by the Department of Transport. The types of road making materials used in the classification are taken directly from this specification, as is some of the nomenclature for concrete pavements, joints and surface markings.

Tolerances in surface levels and finishes have a big impact on plant and labour cost in this type of work. It is consequently important to draw attention to differing or special tolerance requirements by means of additional item description given in accordance with paragraph 5.10.

Many civil engineering employing authorities are concerned with the maintenance and repair of roads as well as with new construction. It should be remembered that the CESMM only gives rules for the measurement of new work in this class and consequently there is no standard method of measuring maintenance and repair operations. There is every reason for measuring such work according to the principals of the CESMM, but the CESMM does not itself include rules for it, neither is the measurement of tying new work into existing work standardized.

The items in class R for kerbs, channels and edgings include backings and beds (rule C3). The items are not said to cover the necessary excavation which is therefore normally measured in class E for earthworks (rule M7). In some cases, convenience to the bill compiler and to the tenderer will be served if excavation for kerbs, channels and edgings is included in the items in class R. A statement in the Preamble in accordance with the note at the foot of page 75 of CESMM2 is necessary to give effect to this arrangement. The same point does not apply to excavation for foundations for traffic sign supports as this is specifically excluded from class E and is covered by rules C4 and M9.

Provision is made in CESMM2 for the measurement of geotextiles associated with road-works (classification R 1 7 0 and rule A3). Note that geotextiles associated with filling are included in class E.

*Schedule of changes in CESMM2*

1. Items for geotextiles are added.
2. Descriptions of reinforcement in concrete pavements are co-ordinated with other classes.
3. Excavation for kerbs and channels may be included in items in this class.

NEWTON TO DINCHOPE ROAD

| Number | Item description | Unit | Quantity | Rate | Amount £ | Amount p |
|---|---|---|---|---|---|---|
| | ROADS AND PAVINGS. | | | | | |
| | Sub-bases, flexible road bases and surfacing. | | | | | |
| R118 | Granular material DTp Specified type 1 depth 475 mm. | m2 | 3760 | | | |
| R216 | Wet-mix macadam DTp Specified clause 808 depth 210 mm. | m2 | 3760 | | | |
| R232 | Dense bitumen macadam DTp Specified clause 903 depth 60 mm. | m2 | 3760 | | | |
| R322 | Rolled asphalt DTp Specified clause 907 depth 40 mm. | m2 | 3760 | | | |
| R341 | Surface dressing depth 25 mm; nominal 20 mm coated chippings. | m2 | 3760 | | | |
| | Precast concrete kerbs, to BS 340 figure 6 bedded and backed with concrete grade C10 cross-section 300 x 350 mm. | | | | | |
| R631 | Straight or curved to radius exceeding 12 m. | m | 917 | | | |
| R632 | Curved to radius not exceeding 12 m. | m | 48 | | | |
| R633 | Quadrants. | nr | 27 | | | |
| | | | PAGE TOTAL | | | |

NEWTON TO DINCHOPE ROAD

| Number | Item description | Unit | Quantity | Rate | Amount £ | p |
|---|---|---|---|---|---|---|
| | ROADS AND PAVINGS. | | | | | |
| R651 | Precast concrete channels to BS 340 figure 8 straight or curved to radius exceeding 12 m; 400 x 150 mm bed. | m | 917 | | | |
| R652 | Precast concrete channels to BS 340 figure 8 curved to radius not exceeding 12 m, 400 x 150 mm bed. | m | 48 | | | |
| R653 | Precast concrete channels to BS 340 figure 8 quadrant, 400 x 150 mm bed. | nr | 27 | | | |
| | Light duty pavements. | | | | | |
| R714 | Granular base depth 150 mm; DTp Specified type 1. | m2 | 1426 | | | |
| R727 | Hardcore base depth 300 mm. | m2 | 387 | | | |
| R774 | In situ concrete grade C15 depth 150 mm. | m2 | 387 | | | |
| R783 | Precast concrete flags to BS 368 type D; thickness 63 mm. | m2 | 1426 | | | |
| R900 | Joint new concrete road to existing concrete road. | m | 27 | | | |
| | | | PAGE TOTAL | | | |

SERVICE ROADS

| Number | Item description | Unit | Quantity | Rate | Amount £ | p |
|---|---|---|---|---|---|---|
| | ROADS AND PAVINGS. | | | | | |
| | Sub-bases, flexible road bases and surfacing. | | | | | |
| R124 | Granular material DTp Specified type 2 depth 150 mm. | m2 | 1039 | | | |
| R170 | Geotextiles; Georam, grade G2. | m2 | 1039 | | | |
| R180 | Additional depth of hardcore. | m3 | 380 | | | |
| | Concrete pavements. | | | | | |
| R414.1 | Carriageway slabs of DTp Specified paving quality concrete depth 150 mm. | m2 | 1039 | | | |
| R414.2 | Carriageway slabs of DTp Specified paving quality concrete depth 150 mm; inclined at an angle exceeding 10°. | m2 | 1764 | | | |
| R443 | Steel fabric reinforcement to BS 4483 nominal mass 3 - 4 kg/m2; type A252. | m2 | 1039 | | | |
| R480 | Waterproof membrane below concrete pavements; 500 grade impermeable plastic sheeting. | m2 | 1039 | | | |
| | Joints in concrete pavements. | | | | | |
| R524 | Expansion joints depth 100 - 150 mm; as detail C drawing 137/51 at 5 m centres. | m | 321 | | | |
| R534 | Contraction joints depth 100 - 150 mm; as detail D drawing 137/51 at 2.5 m centres. | m | 47 | | | |
| | | | | PAGE TOTAL | | |

SERVICE ROADS

| Number | Item description | Unit | Quantity | Rate | Amount £ | p |
|---|---|---|---|---|---|---|
| | ROADS AND PAVINGS. | | | | | |
| | Kerbs, channels and edgings. | | | | | |
| R661 | Precast concrete edgings to BS 340 figure 10 straight or curved to radius exceeding 12 m; 200 x 200 mm concrete grade C10 bed and haunch. | m | 127 | | | |
| R662 | Precast concrete edgings to BS 340 figure 10 curved to radius not exceeding 12 m; 200 x 200 mm concrete grade C10 bed and haunch. | m | 480 | | | |
| | Light duty pavements. | | | | | |
| R713 | Granular base DTp Specified type 2 depth 100 mm. | m2 | 840 | | | |
| R714 | Granular base DTp Specified type 2 depth 150 mm; inclined at an angle exceeding 10° to the horizontal. | m2 | 760 | | | |
| R752.1 | Bitumen macadam DTp Specified clause 912 depth 50 mm. | m2 | 510 | | | |
| R752.2 | Bitumen macadam DTp Specified clause 912 depth 50 mm; inclined at an angle exceeding 10° to the horizontal. | m2 | 37 | | | |
| R782.1 | Precast concrete flags to BS 368 type D depth 50 mm. | m2 | 330 | | | |
| R782.2 | Precast concrete flags to BS 368 type D depth 50 mm; inclined at an angle exceeding 10° to the horizontal. | m2 | 390 | | | |
| | | | | PAGE TOTAL | | |

# *Class S: Rail track*

The measurement of rail track is substantially revised in the second edition of the CESMM. The classification table and accompanying rules are sufficiently different in CESMM2 to justify comprehensive scrutiny. The similarities between the old and the new are relatively few.

The first division now begins with track foundations. These distinguish between bottom ballast and top ballast, terms which are defined in rules D1 and D2. The rates entered against these items should include for the cost of boxing up, trimming to line and level and tamping after the track has been laid.

A new classification at S 2 * * covers taking up existing track. Rule A3 requires the amount of dismantling and arrangements for disposal of the taken-up track to be set out.

The classification at S 3 1–5 0 provides for measurement of lifting, packing and slewing existing track. The items are measured by number but rule A5 requires the length of track to be dealt with to be stated in item descriptions. This means, by the application of paragraph 3.9, that separate items must be given for lifting, packing and slewing of different lengths of track. The length to be stated is defined in rule D3.

The classification at S 4–5 * * covers supplying materials for new track laying. Items for laying itself are included in the classification at S 6 * *. Fittings are separated from sleepers and rails in supplying of plain track but switches and crossings include timbers, fittings and check rails (rule C6).

Classification S 6 * * covers laying new track and does not include supply. Where track laying materials are not supplied by the Contractor, rule A13 requires the arrangements made for supply by the

Employer to be stated. The third division classification for laying track includes items for forming curves (S 6 1–4 2–3). Note that these are not full value items but are measured in addition to the item for plain track itself.

The rules in class S apply to rail track work in industrial and harbour installations as well as to track laid in railway systems proper. Much of the railway work in civil engineering contracts is of the former type.

*Schedule of changes in CESMM2*

Class S has been substantially revised in CESMM2. As there are more changes than there are rules left unaltered, it is impossible to include a schedule of changes. Readers are referred to class S in the CESMM second edition itself.

| Number | Item description | Unit | Quantity | Rate | Amount £ | p |
|---|---|---|---|---|---|---|
| | RAIL TRACK. | | | | | |
| S110 | Track foundation, bottom ballast, crushed granite. | m3 | 636 | | | |
| S120 | Track foundation, top ballast, crushed granite. | m3 | 528 | | | |
| S150 | Track foundation, waterproof membrane, visqueen 1000 gauge. | m2 | 1250 | | | |
| | Taking up track. | | | | | |
| | Welded track on concrete sleepers, fully dismantled and placed in Employer's store at Craven Arms p.w.d. | | | | | |
| S211 | Bull head rail, plain track. | m | 2000 | | | |
| S214 | Bull head rail, turnouts. | nr | 4 | | | |
| S250 | Conductor rails. | m | 4000 | | | |
| S281 | Sundries, buffer stops, approximate weight 2.5 tonnes of steel rail and timber sleeper construction. | nr | 4 | | | |
| S283 | Sundries, wheelstops. | nr | 2 | | | |
| | Lifting, packing and slewing. | | | | | |
| S310 | Bull head rail track, length 20 m, maximum distance of slew 200 mm, maximum lift 100 mm. | nr | 1 | | | |
| | Supplying. | | | | | |
| S425 | Flat bottom rail, reference 113A, mass 56 kg/m. | t | 38 | | | |
| S471 | Sleepers, softwood timber, 2600 x 250 x 130 mm. | nr | 504 | | | |
| | | | | PAGE TOTAL | | |

| Number | Item description | Unit | Quantity | Rate | Amount | |
|---|---|---|---|---|---|---|
| | | | | | £ | p |
| | RAIL TRACK. | | | | | |
| | Supplying. | | | | | |
| S472 | Sleepers, concrete 2600 x 250 x 150 mm type C2. | nr | 27 | | | |
| S481 | Fittings, chairs, type 4. | nr | 1008 | | | |
| S482 | Fittings, base plates, type 3. | nr | 1008 | | | |
| S483 | Pandrol rail fastenings, type 7. | nr | 94 | | | |
| S484 | Plain fish plates, type 9. | nr | 127 | | | |
| S514 | Switches and crossings, turnouts, type T4, drawing 27. | nr | 4 | | | |
| S515 | Switches and crossings, diamond crossing, type DC1 drawing 27. | nr | 2 | | | |
| S581 | Sundries, buffer stops, type B2 approximate weight 2.5 tonnes. | nr | 4 | | | |
| S585 | Sundries, switch heaters, type SH2. | nr | 6 | | | |
| | Laying flat bottom rails. | | | | | |
| S621.1 | Plain track; rail reference 113A, mass 56 kg/m, fish plated joints on timber sleepers. | m | 3200 | | | |
| S621.2 | Plain track; rail reference 113A, mass 56 kg/m, welded joints, on concrete sleepers. | m | 2000 | | | |
| S623 | Form curve in plain track, radius exceeding 300 m, welded joints on concrete sleepers. | m | 500 | | | |
| S624 | Turnouts type T4, drawing 27, length 26.2 m, fish plated joints on timber sleepers. | nr | 4 | | | |
| S625 | Diamond crossings, type DC1, drawing 27, length 27.4 m, fish plated joints on timber sleepers. | nr | 2 | | | |
| | | | PAGE TOTAL | | | |

| Number | Item description | Unit | Quantity | Rate | Amount £ | Amount p |
|---|---|---|---|---|---|---|
| | RAIL TRACK. | | | | | |
| | Laying flat bottom rails. | | | | | |
| S627 | Welded joints by Quick Thermit process. | nr | 10 | | | |
| S681 | Sundries buffer stops approximate weight 2.5 tonnes. | nr | 4 | | | |
| S685 | Sundries switch heaters. | nr | 6 | | | |
| | | | PAGE TOTAL | | | |

# *Class T: Tunnels*

The formation of tunnels and other subterranean cavities is an extreme example of civil engineering work dominated by the characteristics of the ground, as these largely determine the planning, design and construction of such works. For this reason the method of measurement, through the bills of quantities compiled from it, reflects the extent of knowledge (or ignorance) of the ground at the time of tender and provides for alternative methods of working or special expedients that may be required to combat probable difficulties.

However thorough a site investigation there must always remain a measure of surmise in its interpretation of ground type and structure. Moreover, the advancing face of a tunnel is constantly penetrating previously unexplored ground the characteristics of which may only be adequately revealed by the tunnelling operation itself.

Another special factor to be taken account of in the measurement of tunnels is the tolerance of the tunnelling system to variations of the ground. Progress depends on the rate of advance of a limited number of faces. Significant increases in productivity may be achieved by mechanization of the tunnelling process, but mechanization also brings with it a reduced tolerance to variations of the ground—a tolerance which differs from one tunnelling machine to another. In all tunnelling there is a high proportion of time-related cost, the time being affected by variations of the ground and any delays caused by special investigations and treatments of it.

One major decision for any tunnel has to be taken by its designer: should the tunnel design, and particularly the form of ground support, remain constant whatever the local nature of the ground disclosed, or should it be varied depending on the local nature of the ground? This decision can significantly affect the relative importance of some of the items in the Bill of Quantities. For supports of uncertain extent, the items in the bill must be based on an assumed reference condition, representing a best estimate of support requirements. Later variation to the extent of support measured can be related to the ground conditions actually revealed as the tunnel advances. Cost uncertainty arising from these factors is likely to be greater in tunnelling, other than through well-explored and uniform ground, than in other types of work.

For these reasons the rules for measurement of tunnels go further than those for other classes of work in limiting the risk carried by the Contractor. The award of a contract to a contractor who has underestimated the extent and cost of the work to be done is rarely to the benefit of any of the parties involved and is an overall discouragement to development of more economic systems of tunnelling. The incidence of such problems is reduced if the estimation of the extent of support is taken out of the Contractor's area of risk. The principal developments in the CESMM perpetuated in the second edition which move in this direction concern measurement of compressed air working, temporary support and stabilization. Rule A1 requires that all work in this class which is to be carried out under compressed air and within each stated range of air gauge pressure should be measured separately. Items are also to be given for the setting up and operating of plant and equipment for working under compressed air. An allowance is thereby included in the Bill of Quantities for the use of compressed air where it is considered likely to be necessary. The items in the bill are then used to assess payment due to the Contractor when installations for and the use of compressed air are approved by the Engineer.

Rule M8 states that 'Both temporary and permanent support and stabilization shall be measured'. This is not qualified by limiting the work measured to that which is expressly required. As a result the Contractor will be paid separately at the rates he has entered in the bill for the extent of support and stabilization which he provides, whether this is the extent required under the Contract or greater. Alternatively, a Provisional Sum may be given for support and stabilization where the needs for this work are too unpredictable to be priced initially. Whichever alternative is used, the proportion of the financial risk associated with the extent of temporary support and stabilization which may be required is taken by the Employer.

The term 'support and stabilization' includes installing rock bolts, steel arches, erecting timber supports, lagging between arches or timbers, applying sprayed concrete or mesh or link support and grouting. For the purpose of rule M8 forward probing, ground freezing and other ground treatments are also regarded as part of support and stabilization.

Rule M8 is the equivalent of note T13 in the first edition of CESMM. The effect of this note was criticized from some quarters during the life of the first edition. The criticism arose from the Employer's point of view and was in the general form that the rule in note T13 placed too much risk on the Employer and was inconsistent with other classes of the CESMM. However, after careful consideration and a wide collection of opinion, it was decided to retain the principle in CESMM2 and it appears unchanged in rule M8. The decision was based on the view that the risk associated with the ground conditions encountered in tunnels is of exceptional magnitude and justifies exceptional treatment, even when ground investigation is thoroughly conducted before tenders are invited.

The CESMM first edition included items for standing by tunnelling plant whilst support and stabilization work was carried out. These items do not appear in CESMM2 in accordance with the general policy change on standby items discussed earlier.

It is important that a reasonable estimate of the likely quantities of work classed as support and stabilization which will be required is given in the bill. Nominal quantities, which are likely to be exceeded, should not be entered as the possibility of the insertion of excessive rates then arises. Clause 56(2) is unlikely to give any protection to Employers in these circumstances. If work is not undermeasured there is little risk of excessive rates being accepted. A Contractor might be tempted to use inordinate quantities of supports if he is able to price them generously. However, this temptation will invariably be overridden by the larger cost factors which make it in the Contractor's interest that the Works shall be completed without delay. The Contractor will be most unlikely to have an incentive to hold up the main work to do unnecessary support work. The measurement of grouting exemplifies this point. It is not in the interests of the efficient construction of tunnel works that excessive quantities of grout material should be injected at higher pressures than necessary. The method of measurement of grouting by the mass of materials injected set out in the CESMM carries no risk of excessive injection provided that the quantities in the original bill are not seen to be underestimated.

A further safeguard exists in the duty which the Engineer has under the Contract through his Representative to supervise the construction of the Works. He has the authority to issue instructions regarding the manner of execution of the Works (clause 13). The Engineer, his Representative and their staffs should involve themselves, by positive contribution as well as by passive approval, in the planning and design of the Temporary Works for and methods of construction of tunnels.

The measurement rules for tunnels in the CESMM in these respects give additional protection to contractors from unforeseen risks and thereby reduce the claims which would otherwise arise under clause 12. This arrangement works to the benefit of the Employer and the Contractor. Where the Engineer is experienced in tunnelling techniques and knowledgeable about costs and the Contractor is ready to comply with the spirit as well as the letter of his contractual obligations the work can be controlled on the Employer's behalf to a greater extent than otherwise.

The measurement rules for tunnelling work are straightforward in most other respects and are little changed in the second edition of the CESMM. It is recommended that tunnelling and other subterranean work should be fully divided into separate bill parts (in accordance with paragraph 5.8) because of the heavy dependence of tunnelling costs on location and construction method.

Excavation of tunnels and shafts (items T 1 * *) is separated from lining (items T 2–7 * *). A distinction is drawn between rock and soft material, and an excavated surface area measurement is given for trimming and mucking out overbreak and also for back grouting of linings to fill voids where this is required.

The term 'pressure grouting' is used in the CESMM to identify treatment intended to support and stabilize the ground around a tunnel. This should not be confused with 'back grouting' to fill voids caused by overbreak. Back grouting to fill voids is not measured; it is allowed for in the prices for excavated surface areas. Rule A5 requires details to be given of the filling required for voids due to overbreak. Rule D2 defines the diameter to be stated for excavation items as the external diameter. The diameter stated for linings is normally the external diameter for primary linings and the internal diameter for secondary linings. Item descriptions should make this clear. Where a breakaway from a shaft or tunnel involves breaking out linings, paragraph 5.18 leads to no measurement of the removed lining being made.

The calculation of quantities for volumes of excavation, areas of excavated surfaces and volumes of in situ linings are based on payment lines shown on the Drawings. Any cavity formed outside such payment lines is deemed to be overbreak. This is the effect of rules M2, M4 and M5. If no payment lines are shown on the Drawings overbreak is assumed to begin either at the limit of the Permanent Works to be constructed in the tunnel or shaft or at the minimum specified size of the void required to be created to accommodate the Permanent Works. Payment lines should not be regarded as indicating an expected volume of overbreak. They are lines of payment convention, not lines to show limits of excavation required by the Contract.

The second edition of the CESMM provides for the disposal of excavated materials arising from tunnelling operations in rules C1 and A4. Materials arising from tunnelling operations are to be taken into account in calculating volumes of filling, where the material is suitable in accordance with rule M19 of class E.

CESMM2 refers to other classes for rules governing calculation of quantities for in situ concrete lining (class F—rules M1 and M2), steel arches (class M—rules M2–M6) and timber (class O—rule M1).

Rules D3 and D4 establish that reinforcement of various types and steel fabric used as an internal support in tunnels are not classed as concrete reinforcement.

The subject of contract practices in tunnelling has been considered by a working party of CIRIA. Its report* includes detailed commentary on the application of the first edition of the CESMM to tunnel-

* Construction Industry Research and Information Association. *Tunnelling—improved contract practices*. London, 1978, Report 79.

ling work. It recommends the use of a referencing system for ground conditions involving grouping of the expected conditions with respect to their varying effect on tunnelling methods and costs. Where this system is used, the group definitions should be referred to in the Preamble statement required by paragraph 5.5 of the CESMM and items for excavation and excavated surfaces should be given separately for each group. This is illustrated by items T128.1 and T128.2 in the example bill.

*Schedule of changes in CESMM2*

1. Locations for disposal of excavated material are to be stated if on the Site.
2. Standby items are deleted.
3. A minimum volume of $0{\cdot}25\ m^3$ for rock excavation is introduced.
4. Rules are added for calculating quantities of steel arches and in situ concrete linings.
5. A note states that in situ concrete tunnel lining may be measured using the rules in other classes if complex shapes are required.

| Number | Item description | Unit | Quantity | Rate | Amount £ | p |
|---|---|---|---|---|---|---|
| | FOREBAY TUNNEL AND OVERFLOW SHAFT. | | | | | |
| | Tunnel excavation diameter 2.5 m. | | | | | |
| T112.1 | In rock; straight. | m3 | 2180 | | | |
| T112.2 | In rock; curved, material to be used as filling. | m3 | 520 | | | |
| T132 | Shaft excavation diameter 2.8 m in rock; straight. | m3 | 280 | | | |
| T170 | Excavated surfaces in rock; voids filled with cement grout as Specification clause 137/T17. | m2 | 4352 | | | |
| T232.1 | In situ reinforced cast concrete primary straight tunnel lining internal diameter 2.0 m; concrete as Specification clause 137/F3. | m2 | 820 | | | |
| T232.2 | In situ reinforced concrete primary curved tunnel lining internal diameter 2.0 m; concrete as Specification clause 137/F3. | m3 | 175 | | | |
| T252.1 | In situ straight lining formwork finish grade T2 internal diameter 2.0 m. | m2 | 3566 | | | |
| T252.2 | In situ curved lining formwork finish grade T2 internal diameter 2.0 m. | m2 | 760 | | | |
| | | | PAGE TOTAL | | | |

| Number | Item description | Unit | Quantity | Rate | Amount £ | p |
|---|---|---|---|---|---|---|
| | DIVERSION TUNNEL. | | | | | |
| | Work to be executed under compressed air at gauge pressure not exceeding one bar. | | | | | |
| T128.1 | Excavation straight tunnel diameter 10.3 m in Group 5 material. | m3 | 41000 | | | |
| T128.2 | Excavation straight tunnel diameter 10.3 m in Group 6 material. | m3 | 8800 | | | |
| T180.1 | Excavated surface Group 5 material; voids filled with cement grout as Specification Clause 137/T17. | m2 | 15800 | | | |
| T180.2 | Excavated surface in Group 6 material; voids filled with cement grout as Specification Clause 137/T17. | m2 | 3400 | | | |
| T538 | Cast iron bolted segmental tunnel lining rings external diameter 10.3 m; nominal width 450 mm comprising 16 segments maximum piece weight 110 kg 136 bolts and grummets and 272 washers. | nr | 1330 | | | |
| T571 | Parallel circumferential packing for preformed segmental tunnel linings; bitumen impregnated fibreboard thickness 8 mm. | nr | 1329 | | | |
| T574 | Leaf fibre caulking for preformed segmental tunnel linings. | m | 51500 | | | |
| | | | PAGE TOTAL | | | |

| Number | Item description | Unit | Quantity | Rate | Amount £ | p |
|---|---|---|---|---|---|---|
| | FOREBAY TUNNEL AND OVERFLOW SHAFT. | | | | | |
| | In situ lining to shafts. | | | | | |
| T332 | Cast concrete primary internal diameter 2.3 m; concrete as Specification clause 137/F3. | m3 | 130 | | | |
| T352 | Formwork finish Grade T1 diameter 2.3 m. | m2 | 615 | | | |
| | Support and stabilization. | | | | | |
| T811 | Rock bolts impact expanding mechanical 55 mm diameter with 30 mm square shank maximum length 5 m. | m | 108 | | | |
| | Pressure grouting. | | | | | |
| T831 | Sets of drilling and grouting plant. | nr | 1 | | | |
| T832 | Face packers. | nr | 180 | | | |
| T834 | Drilling and flushing diameter 50 mm length 5 - 10 m. | m | 1080 | | | |
| T835 | Re-drilling and flushing holes length 5 - 10 m. | m | 540 | | | |
| T836 | Injection of cement grout as Specification clause 137/T801. | t | 85 | | | |
| T840 | Forward probing length 10- 15 m. | m | 480 | | | |
| | | | PAGE TOTAL | | | |

# *Class U: Brickwork, blockwork and masonry*

The class dealing with brickwork, blockwork and masonry has to cover a very wide range of uses of these materials: from massive masonry in a breakwater or dock wall to isolated panels of ordinary house bricks which might be specified in the buildings incidental to a work of civil engineering. The measurement rules attempt to encompass this range by aiming at the middle. The result is that the rules are somewhat less detailed than those for brickwork and blockwork which are used in building measurement practice.

This class has been considerably simplified in the classification table of the second edition of the CESMM, although the actual items measured are little different. The item descriptions generated by the classification table in this class must be considerably expanded to produce complete descriptions as required by rules A1–A5.

In CESMM2 brickwork and blockwork walls are no longer described as being 'one-brick' or 'one-block' thick. It is now only necessary to state the dimensions of the bricks, blocks or stones in accordance with rule A1 and the thickness of the wall in accordance with rule A5. All walls up to 1 m thick are measured by the square metre. Walls exceeding 1 m thick are measured by the cubic metre with the thickness stated (rule A5).

The rules for cavity and composite walls are rules M1, M6 and A4. They require that each of the two skins of such walls shall be measured and that they should be identified in item descriptions. The area measured for ties between skins is the area tied. In the case of a cavity wall this is the area of the wall constructed, which is half the area of the two skins.

Where a brick or block wall is tied to a concrete wall the ties would normally be included in the items in class U. Inserts must also be measured in class G. In accordance with rule C7 of class G, it should be stated that only fixing of the inserts is included in the class G item when the supply of the ties is measured in class U.

Brickwork, blockwork and masonry which are in straightforward geometrically simple shapes are measured very simply using the CESMM rules. Where there are many different surface features which interfere with simple geometry, some measurement complications can arise. Class U generally deals with surface features as separate items (U * 7 *) which are priced to cover the extra cost of forming the surface features. The rules for deductions from areas and volumes have been amended in the CESMM second edition such that areas of surface features are ignored in calculations of volumes and areas (rule M2) and the determination of thicknesses (rule D3). The only deductions now made from volumes and areas of walls are for holes and openings exceeding 0·25 $m^2$ in cross-sectional area (rule M2). An example is that the area (or volume) measured for a wall includes the areas of any plinths and cornices. The effect of rule A7 is that the item descriptions for surface features do not need to include the size unless the cross-sectional area exceeds 0·05 $m^2$. This is the area, for example, of a cill or coping of cross-section 224 mm square or 250 mm × 200 mm. Rule M2 specifically excludes the measurement of the areas of cills and copings from the areas measured for walls. This is why the material of which cills and copings are made is to be stated in the descriptions of their own items (U * 7 1). This is a change from the CESMM first edition in which copings and cills were measured effectively 'extra over' in common with other surface features. Another change in CESMM2 is that columns and piers are now measured by height instead of by area.

Surface features in this class of work must always be considered carefully when the bill is in preparation. The first impression gained by the taker-off of the significance of the shapes to be measured may not coincide with another person's impression. An obvious and extreme example of this type of difficulty is illustrated by the wall with the cross-sectional profile shown in Fig. 27. This could be classed as a wall 900 mm thick with 200 mm × 200 mm projections or as a wall 1100 mm thick with 200 mm × 200 mm rebates. If it is the former it should be measured by area (U * 4 *); if it is the latter it should be measured by volume (U * 5 *) with no deduction for the

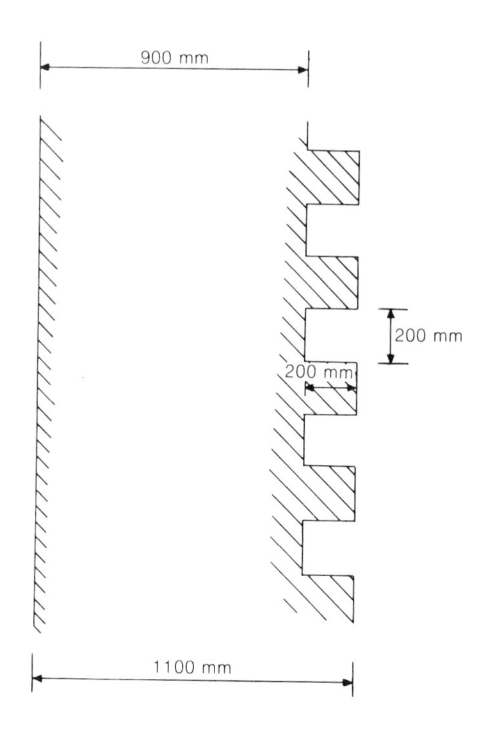


Fig. 27. A wall with these surface features could be considered as of thickness 900 mm with 200 mm × 200 mm projections or as of thickness 1100 mm with 200 mm × 200 mm rebates. Additional description must be given when work such as this—straddling the boundary of two parts of the Work Classification—is to be measured. This is the effect of paragraph 5.13. The wall in the figure should be identified using additional description to avoid uncertainty

volume of the rebates. If the overall thickness had been 1300 mm instead of 1100 mm, the volume could have been based on the overall thickness with no deduction for the rebates, or on the root thickness with no addition for the projections. It is important that each particular bill should be made free of such ambiguities by the use of clear additional item description. In this way tenderers can be sure to price on the same assumption as was made in the calculation of the quantities.

In the same context, fair facing as a surface feature (U * 7 8) is always measured in square metres quite independently of whether the material which is being fair faced is facing or common brickwork or blockwork. 'Facing brickwork' (U 2 * *) in the first division of classification refers to the material; 'fair facing' in the third division refers to the additional labour cost of laying any material to a fair face.

Among the ancillaries are items for building in pipes and ducts. In CESMM2 these are grouped into those not exceeding 0·05 $m^2$ in cross-sectional area (250 mm diameter) and those which are larger. Lengths of pipes built in have to be stated if they are more than 1 m long (rule A10). As usual, if the pipes are supplied under another item, descriptions must say so (rule C2).

*Schedule of changes in CESMM2*

1. The classification tables in the second division have been simplified.
2. The classification table in the third division now provides for curved battered walls, vertical and battered facing to concrete and casings to metal sections.
3. The classification table in the third division now provides for the measurement of cavity ties and tying to other materials.
4. Items for building in pipes and ducts have been amended.
5. Measurement of centring to arches has been deleted.
6. Columns and piers are measured by height.
7. Copings and cills are measured full value.

| Number | Item description | Unit | Quantity | Rate | Amount | |
|---|---|---|---|---|---|---|
| | BRICKWORK, BLOCKWORK AND MASONRY. | | | | | |
| | Common brickwork to BS 3921; stretcher bond, flush pointed mortar type M2. | | | | | |
| U111 | 102.5 mm nominal thickness vertical straight walls; cavity construction. | m2 | 287 | | | |
| U160 | Columns and piers, cross sectional dimensions 600 x 600 mm. | m | 43 | | | |
| | Ancillaries. | | | | | |
| U182.1 | Damp proof courses; width 100 mm to BS 743 type D. | m | 150 | | | |
| U182.2 | Damp proof courses; width 225 mm to BS 743 type D. | m | 17 | | | |
| U185 | Concrete infill grade 1; thickness 50 mm. | m2 | 31 | | | |
| U186 | Fixing and ties; galvanised mild steel in accordance with Specification clause 5/27. | m2 | 304 | | | |
| U187 | Built-in pipes and ducts cross-sectional area not exceeding 0.05 m2 | nr | 69 | | | |
| U188 | Built-in pipes and ducts cross-sectional area 0.10 m2. | nr | 6 | | | |
| | | | PAGE TOTAL | | | |

| Number | Item description | Unit | Quantity | Rate | Amount | |
|---|---|---|---|---|---|---|
| | BRICKWORK, BLOCKWORK AND MASONRY. | | | | | |
| | Facing brickwork. | | | | | |
| | Plowden Red facings as Specification clause 5/29. | | | | | |
| U211 | 102.5 mm nominal thickness vertical straight wall; stretcher bond flush pointed, mortar type M3 cavity construction. | m2 | 321 | | | |
| | Surface features. | | | | | |
| U271.1 | Brick on edge coping. | m | 143 | | | |
| U271.2 | Special sill as detail D drawing 137/97. | m | 37 | | | |
| | Brickwork ancillaries. | | | | | |
| U282 | Damp proof courses; width 100 mm to BS 743 type D. | m | 187 | | | |
| U287 | Built-in pipes and ducts cross-sectional area not exceeding 0.05 m2 | nr | 69 | | | |
| U288 | Built-in pipes and ducts cross-sectional area 0.10 m2. | nr | 6 | | | |
| | | | PAGE TOTAL | | | |

| Number | Item description | Unit | Quantity | Rate | Amount | |
|---|---|---|---|---|---|---|
| | BRICKWORK, BLOCKWORK AND MASONRY. | | | | | |
| | Engineering brickwork class B to BS 3921, stretcher bond, flush pointed mortar type M4. | | | | | |
| U333 | 450 mm nominal thickness battered straight wall. | m2 | 87 | | | |
| U334 | 450 mm nominal thickness battered curved wall. | m2 | 184 | | | |
| U371 | Bullnosed copings. | m | 47 | | | |
| U375 | Bullnosed corbels. | m | 83 | | | |
| U383 | Movement joint; 25 mm bitumen impregnated fibreboard, mean width 450 mm with plastijoint sealer both sides. | m | 27 | | | |
| U384 | Bond to existing work. | m2 | 14 | | | |
| | Lightweight blockwork, hollow block to BS 6073, stretcher bond, flush pointed mortar type M1. | | | | | |
| U411.1 | Vertical 100 mm nominal thickness straight wall. | m2 | 203 | | | |
| U411.2 | Vertical 140 mm nominal thickness straight wall. | m2 | 127 | | | |
| U482.1 | Damp proof courses; width 100 mm to BS 743 type D. | m | 83 | | | |
| U482.2 | Damp proof courses; width 140 mm to BS 743 type D. | m | 46 | | | |
| | | | PAGE TOTAL | | | |

| Number | Item description | Unit | Quantity | Rate | Amount | |
|---|---|---|---|---|---|---|
| | BRICKWORK, BLOCKWORK AND MASONRY. | | | | | |
| | Dense concrete blockwork, solid block to BS 6073, stretcher bond, flush pointed mortar type M1. | | | | | |
| U511 | Vertical 140 mm nominal thickness straight wall. | m2 | 83 | | | |
| U582 | Damp proof courses; width 140 mm to BS 743 type D. | m | 27 | | | |
| U587 | Built-in pipes and ducts cross-sectional area not exceeding 0.05 m2. | nr | 41 | | | |
| | Ashlar masonry Portland stone rubbed finish Specification clause 5/40 mortar type M2. | | | | | |
| U735 | 300 mm nominal thickness vertical facing to concrete flush pointed. | m2 | 4387 | | | |
| U771 | Surface features; rounded copings 1200 x 600 mm as drawing 137/91. | m | 1236 | | | |
| U786 | Ancillaries fixings and ties; Alleyslots at 1500 mm centres. | m2 | 4387 | | | |
| U787 | Ancillaries; built-in pipes and ducts cross-sectional area 0.025 - 0.25 m2. | nr | 45 | | | |
| U799 | Bollards diameter 600 mm height 900 mm as drawing 137/93. | nr | 24 | | | |
| | | | PAGE TOTAL | | | |

# *Classes V and W: Painting and waterproofing*

Classes V and W are considered together because the rules which they contain have much in common. They have been changed hardly at all in the CESMM second edition.

The includes and excludes list for class V establishes some important boundaries of item coverage for painting. The class effectively includes surface treatments to all work which are carried out on the Site. Treatments which are to be carried out before delivery to the Site will be deemed to be covered by the items in the other classes which embody the supply of the various materials and components to the Site. Only in structural metalwork (class M) are separate items given for predelivery surface treatments; in all other classes the specified surface treatment is deemed to be included in the items.

The first division of classes V and W and rule A1 establish that item descriptions for painting and waterproofing should clearly identify the materials to be used and their method of application. Any specified prior preparation of surfaces is deemed to be included (rule C1) and surface preparation prior to painting will be deemed to be covered by the prices entered against the items for painting. It is consequently important to identify the preparation in any case in which there is more than one specified surface preparation requirement for the same base material or for the same surface treatment applied to it.

Labours for painting and waterproofing are not given separate items. This assumes that drawings showing details of joints, laps and edges are supplied to tenderers. Consequently the prices for the areas and for the lengths of narrow widths must allow for all labours. Most surfaces of any magnitude are measured by area. The excep-

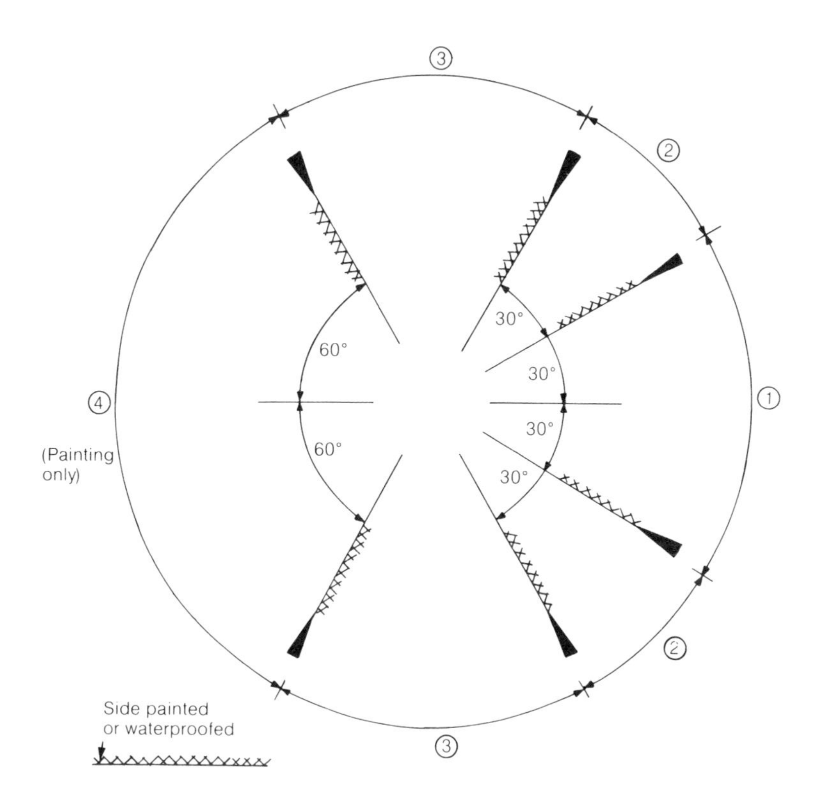


Fig. 28. Inclination zones used for classification of painting and waterproofing of plane surfaces of width exceeding 1 m as given in the third division of classes V and W. (See also rule M2.) Note the precise boundaries of the zones. For example, a soffit surface inclined at 60° to the horizontal is classed in zone 4; one inclined at 61° is classed in zone 3

tions are surfaces one metre wide or less which are measured by length and isolated groups of surfaces which are measured by number. Painting of metal sections and pipes is measured by area irrespective of width.

The facility for defining 'isolated groups of surfaces' is a shorthand method of identifying particular components to be painted or waterproofed for which it would be tedious or unhelpful to give separate items for each of their separate surfaces. Sumps are a common example. An extreme example—to make the point clearly—might be the requirement to paint the bust of the chairman of the reservoir committee which is to be placed at a gate. It would be difficult and unhelpful to attempt to distinguish the width or inclination of all the separate surfaces making up his or her craggy likeness. It would be easier, if less respectful, to itemize the work under the heading 'Isolated groups of surfaces' as 'Paint bust of committee chairman at the gate—sum'. If there were several identical busts to be painted they would be enumerated, 'sum' being equivalent to 'nr 1'.

The area measured for painting and waterproofing is generally the net area covered, subject to the normal inclusion of holes less than 0·5 $m^2$ in area. The area measured for painting metal sections and pipework is slightly different in that the prices should include an allowance for additional areas of valves and joints in pipework and for connections, rivet and bolt heads in metal sections. This is the effect of rules M4, M5, C2 and C3 of class V. Lagged pipes should still be regarded as pipes for the purposes of measurement.

The various inclination boundaries and zones for surfaces exceeding 1 m wide are specified in the third division of both classification tables. They are all shown in Fig. 28. The zone for soffit and lower surfaces in painting (V * * 4) is not used in waterproofing on the assumption that lower surfaces are seldom waterproofed. Feature 4 in the third division of waterproofing is allocated to curved surfaces of any inclination.

### *Schedule of changes in CESMM2*

Only minor textual changes.

| Number | Item description | Unit | Quantity | Rate | Amount | |
|---|---|---|---|---|---|---|
| | PAINTING. | | | | | |
| V113 | Zinc rich primer paint on metal surfaces other than metal sections and pipework inclined at an angle exceeding 60° to the horizontal; in one coat. | m2 | 47 | | | |
| V116 | Zinc rich primer paint on metal surfaces other than metal sections and pipework width not exceeding 300 mm; in one coat. | m | 91 | | | |
| V118 | Zinc rich primer paint on metal surfaces other than metal sections and pipework isolated groups of surfaces; 300 x 200 mm inspection covers and frames in one coat; | nr | 14 | | | |
| | Oil paint in three coats. | | | | | |
| V313 | On metal surfaces other than metal sections and pipework; inclined at an angle exceeding 60° to the horizontal. | m2 | 47 | | | |
| V316 | On metal surfaces other than metal sections and pipework of width not exceeding 300 mm. | m | 91 | | | |
| V318 | On metal surfaces other than metal sections and pipework isolated groups of surfaces; 300 x 200 mm inspection covers and frames; | nr | 14 | | | |
| V326 | Timber surfaces of width not exceeding 300 mm. | m | 14 | | | |
| | | | PAGE TOTAL | | | |

| Number | Item description | Unit | Quantity | Rate | Amount | |
|---|---|---|---|---|---|---|
| | PAINTING. | | | | | |
| | Emulsion paint in three coats. | | | | | |
| V533 | Smooth concrete surfaces inclined at an angle exceeding $60^{\circ}$ to the horizontal. | m2 | 487 | | | |
| V534 | Smooth concrete soffit and lower surfaces inclined at an angle not exceeding $60^{\circ}$ to the horizontal. | m2 | 326 | | | |
| V563 | Brickwork and blockwork surfaces inclined at an angle exceeding $60^{\circ}$ to the horizontal. | m2 | 3867 | | | |
| V566 | Brickwork and blockwork surfaces of width not exceeding 300 mm. | m | 187 | | | |
| V633 | Cement paint on smooth concrete surfaces inclined at an angle exceeding $60^{\circ}$ to the horizontal; in one coat. | m2 | 27 | | | |
| V636 | Cement paint on smooth concrete surfaces of width not exceeding 300 mm; in one coat. | m | 14 | | | |
| | | | PAGE TOTAL | | | |

GENERATOR HOUSE

| Number | Item description | Unit | Quantity | Rate | Amount | |
|---|---|---|---|---|---|---|
| | WATERPROOFING. | | | | | |
| W131 | Damp proofing, waterproof sheeting to upper surfaces inclined at an angle not exceeding $30^{\circ}$ to the horizontal; polythene to BS 3012 in one layer thickness 5 mm. | m2 | 320 | | | |
| W211 | Tanking, asphalt to upper surfaces inclined at an angle not exceeding $30^{\circ}$ to the horizontal; mastic asphalt to BS 1097 in one coating thickness 25 mm. | m2 | 410 | | | |
| W213 | Tanking, asphalt to surfaces inclined at an angle exceeding $60^{\circ}$ to the horizontal; mastic asphalt to BS 1097 in one coating thickness 25 mm. | m2 | 534 | | | |
| W311 | Roofing asphalt upper surfaces inclined at an angle not exceeding $30^{\circ}$ to the horizontal; mastic asphalt to BS 988 in two coatings total thickness 25 mm including heavy gauge polythene isolating membrane non-staining roofing felt to BS 747 type 1C expanded polystyrene thickness 25 mm and concrete grade 10 screed average thickness 75 mm. | m2 | 450 | | | |
| W316 | Roofing asphalt surfaces of width not exceeding 300 mm; mastic asphalt to BS 988 in two coatings total thickness 25 mm. | m | 726 | | | |
| W441 | Protective layer sand and cement screed type A thickness 25 mm upper surfaces inclined at an angle not exceeding $30^{\circ}$ to the horizontal. | m2 | 410 | | | |
| | | PAGE TOTAL | | | | |

PUMP HOUSE

| Number | Item description | Unit | Quantity | Rate | Amount | |
|---|---|---|---|---|---|---|
| | WATERPROOFING. | | | | | |
| | Roofing. | | | | | |
| W321 | Profiled aluminium sheet to BS 4868 type A 1 mm thick in one layer upper surfaces inclined at an angle not exceeding 30° to the horizontal. | m2 | 463 | | | |
| W323 | Profiled alumimium sheet to BS 4868 type A 1 mm thick in one layer surfaces inclined at an angle exceeding 60° to the horizontal. | m2 | 125 | | | |
| W331 | Waterproof sheeting upper surfaces inclined at an angle not exceeding 30° to the horizontal; corrugated plastic translucent sheet to BS 4154 1.5 mm thick. | m2 | 57 | | | |
| | | | PAGE TOTAL | | | |

# *Class X: Miscellaneous work*

Class X contains three types of work: fencing, drainage to structures above ground and rock filled gabions.

The classification table at X 1–2 * * provides the rules of measurement for fences, gates and stiles. Rules A2 and A3 require the types and principal dimensions of fences, gates and stiles to be stated. The measurement unit for fences is the metre. In accordance with paragraph 5.18 and rule M1 this means that the net actual length is measured. For a zigzag fence this is the developed length. In CESMM2, coverage rule C1 makes it clear that the items for fencing include all work associated with foundations for fence and gate posts.

Drainage for structures above ground is covered by the classification table at X 3 * *. This measurement is also based on a simple item description for the components and classification by the material and the nature of the components. Rule D3 defines these components which are to be dealt with as fittings. Other components, such as holder bats and brackets, are not defined as fittings. They are therefore classed as supports which are deemed to be included in the main items (rule C3).

Rock filled gabions are included in class X as an addition in the second edition of the CESMM. They are covered at classification X 4 * *. Box and mattress gabions are given separately, the former measured by number the latter by square metre. Rule D4 states that the boundary between the two has a thickness of 300 mm. Rule A5 requires additional description to be given for the filling and wire mesh. Any express requirements about the source of the filling should be given as further additional description.

*Schedule of changes in CESMM2*

Items are added for rock filled gabions.

| Number | Item description | Unit | Quantity | Rate | Amount | |
|---|---|---|---|---|---|---|
| | MISCELLANEOUS WORK. | | | | | |
| | Fences. | | | | | |
| X113 | Timber post and rail fence height 1.3 m; to BS 1722 Part 7. | m | 187 | | | |
| X133 | Concrete post and wire plastic coated chain link fence to BS 1722 Part 1 height 1.4 m concrete grade 10 foundations 450 x 450 x 450 mm deep. | m | 470 | | | |
| X136 | Concrete post and wire anti-intruder chain link fence to BS 1722 Part 10 height 2.9 m; with cranked posts concrete grade 10 foundations 600 x 600 x 750 mm deep. | m | 305 | | | |
| X163.1 | Timber close boarded fence height 1.4 m; with concrete posts to BS 1722 Part 5 concrete grade 10 foundations 300 x 300 x 450 mm deep. | m | 27 | | | |
| X163.2 | Timber close boarded fence height 1.4 m; with concrete posts to BS 1722 Part 5 concrete grade 10 foundations on surface inclined at an angle exceeding 10°. | m | 13 | | | |
| X191 | Chestnut pale fence height 900 mm; with timber posts to BS 1722 Part 4. | m | 403 | | | |
| X193.1 | Mild steel continuous bar fence height 1.372 m; to BS 1722 Part 8 concrete grade 10 foundations 450 x 450 x 450 mm deep. | m | 1871 | | | |
| X193.2 | Mild steel unclimbable fence height 1.372 m; to BS 1722 Part 9 concrete grade 10 foundations 450 x 450 x 450 mm deep. | m | 108 | | | |
| | | | PAGE TOTAL | | | |

| Number | Item description | Unit | Quantity | Rate | Amount | |
|---|---|---|---|---|---|---|
| | MISCELLANEOUS WORK. | | | | | |
| | Gates and stiles. | | | | | |
| X215 | Timber field gate width 3.353 m height 1.143 m; to BS 3470. | nr | 14 | | | |
| X235 | Metal field gate width 3.353 m height 1.143 m; to BS 3470. | nr | 2 | | | |
| X295 | Entrance gate width 3.5 m height 1.4 m; galvanised mild steel angle sections filled in with plastic coated chain link fencing as detail D drawing 137/75. | nr | 1 | | | |
| | Drainage to structures above ground unplasticized PVC to BS 4576. | | | | | |
| X331.1 | Gutters; 100 mm diameter. | m | 47 | | | |
| X331.2 | Gutters; 150 mm diameter. | m | 36 | | | |
| X332.1 | Gutter bends 90$^{o}$; 100 mm diameter. | nr | 10 | | | |
| X332.2 | Gutter outlets; 150 mm diameter. | nr | 8 | | | |
| X333.1 | Downpipes; 100 mm diameter. | m | 23 | | | |
| X333.2 | Downpipes; 150 mm diameter. | m | 16 | | | |
| X334.1 | Swan necks; 100 mm diameter. | nr | 4 | | | |
| X334.2 | Shoes; 150 mm diameter. | nr | 3 | | | |
| | | | PAGE TOTAL | | | |

| Number | Item description | Unit | Quantity | Rate | Amount | |
|---|---|---|---|---|---|---|
| | MISCELLANEOUS WORK. | | | | | |
| | Rock filled gabions. | | | | | |
| X410.1 | Box gabion 1 x 1 x 2 m: 4 mm x 100 x 100 mm galvanised wire mesh, excavated rock grade GB1. | nr | 27 | | | |
| X410.2 | Box gabion 1 x 1 x 2 m; 4 mm x 100 x 100 mm galvanised wire mesh, excavated rock grade GB1. | nr | 45 | | | |
| X420 | Mattress gabion, thickness 150 mm, 4 mm x 100 x 100 mm galvanised wire mesh, excavated rock grade GB2. | m2 | 50 | | | |
| | | | PAGE TOTAL | | | |

# *Class Y: Sewer renovation and ancillary works*

Rules of measurement for sewer renovation and ancillary works are included in the second edition of the CESMM for the first time as a response to the growing volume of this type of civil engineering work. Class Y differs from the other CESMM classes in that it is for measuring repairs and alterations to existing structures, not new work. Possible exceptions to this principle in class Y are the replacement of small areas of defective work at Y 1 5 *, the replacement of flap valves at Y 4 2 * and the construction of new manholes at Y 5–6 * *. It is important therefore to define clearly any boundaries between new and existing sewers in the contract documents.

Temporary Works may be a substantial element of sewer renovation work. The General items section of the CESMM should be used fully, paying particular attention to Specified Requirements and Method-Related Charges. Specified Requirements might be used in connection with such work as closed circuit television surveys, proving sewer dimensions, core sampling and any other work which is expressly required but which does not form part of the Permanent Works. Provision should also be made for the Contractor to include Method-Related Charges for such work as lead-in trenches, temporary access shafts, pumping and diversion works.

The special circumstances of sewer renovation and ancillary works preclude simple substitution of rules from classes I–L into class Y. Only a few rules from these classes have been incorporated either in their original or a modified form. Those rules which have not been modified have been repeated in full in class Y to avoid ambiguity and so that class Y is a complete set of rules. It can be used for a sewer renovation contract with only general items added to make a complete bill of quantities.

Rules A2 and A5 require the features of the main sewer to be stated. Rules A3 and A4 relate to the distinction between 'man entry' and 'non-man-entry' sewers. They are only invoked when the Engineer has decided to restrict the Contractor's choice of methods. Rule A3 deals with methods of working from within sewers, whether by remote control as in 'non-man-entry' systems or manually as in 'man entry' systems. The decision to carry out the works from above by excavation (A4) might be taken as a result of the sewer being 'non man entry' or for other reasons such as difficult access or poor safety considerations.

The classification table divides the main work into preparation, stabilization, renovation and manholes. In addition there are supplementary classifications for work to laterals undertaken after renovation works are complete and for work which is interrupted.

The preparation classification (Y 1 * *) covers preparatory work such as removal of silt, grease, encrustation and tree roots which is done before sewer renovation. The CESMM assumes that the condition of existing sewers can be ascertained before establishing quantities for stabilization and renovation. Preparation work often forms a preliminary contract as it is difficult to ascertain the nature and extent of the renovation and ancillary works required until the sewer has been cleaned. The measurement of cleaning at classification Y 1 1 0 is very simple. Differing cleaning requirements and standards must be identified for the different locations billed in accordance with rule A1. The items are deemed to include making good any damage caused by cleaning (rule C2) but not any damage which is revealed as a result of the cleaning. Classification Y 1 2 * covers removal of intrusions. An intrusion is a projection into the bore of a sewer. Artificial intrusions may include isolated projecting bricks, projecting rubber O rings and dead services. Laterals are any drains or sewers which are connected to the sewer being renovated. They are prepared by cleaning (Y 1 1 0) or by sealing (Y 1 3–4 *). Sealing laterals and other pipes is measured in cubic metres. To calculate either, the cross-sectional dimensions and length must be known or a conversion factor must be agreed between the Engineer and the Contractor converting cubic metres to 'bags of grout'. Local internal repairs are repairs to the structural fabric of the sewer which are carried out from within the sewer. Examples are isolated patch repairs, repairs to bellmouths and Y junctions which are not to be renovated and repairs around laterals.

Stabilization of existing sewers is carried out by pointing, joint sealing and external grouting. The area measured for pointing at classification Y 2 1 0 excludes areas which have been repaired and measured using classification Y 1 5 *. Different types of pointing such as hand pointing and pressure pointing are distinguished by the different locations identified in accordance with rule A1. Pipe joint sealing measured at classification Y 2 2 0 covers both joint sealing and repairs to cracked joints whether longitudinal or circumferential.

Renovation of sewers itself is measured at classification Y 3 * *. It covers the methods by which the performance of a length of sewer is improved by incorporating the original sewer fabric but excluding maintenance operations such as root or silt removal and local internal repairs. Renovation may be carried out using a variety of techniques which are listed in the second division at classification Y 3 1–5 *. Sliplining is a method in which lengths of pipe lining are jointed before being moved into their permanent positions. The two types of sliplining given in the third division are indicative rather than definitive and are qualified by rule A8 which requires greater detail to be stated. Where a copolymer lining is specified the item code Y 3 1 9 would be appropriate. In situ jointed pipe lining is a method in which lengths of pipe lining are jointed at their permanent positions. Segmental linings are circular or non-circular sewer linings which are made up from pairs of upper and lower segments which are jointed near their springings. There is a limited number of proprietary sewer renovation systems. All proprietary systems fall within classification Y 3 4 *. Rule A8 distinguishes them in detail. Rule A9 requires curved in situ jointed and segmental linings to be identified. Other renovation techniques are either impossible to install to a curve or are priced similarly for curved and straight work. Curved work is defined as work on sewers whose offset exceeds 35 mm per metre.

Engineers sometimes allow the Contractor to choose an appropriate renovation technique, usually from a number of permitted options. Where this approach is used, a reference must be made in the Preamble using the procedure for Contractor selection of alternatives referred to in paragraph 5.4 of the CESMM.

Classification Y 3 6 0 is used to measure grouting of annular voids as defined in rule D4. Grouting is measured by volume. This may necessitate adoption of a measurement convention or conversion factor relating to bags of grout or the volume of material passing through the pump. Annulus grouting, as well as filling the annular void, consolidates the brickwork, fills the cracks and, in certain circumstances, may be used to fill voids which have formed outside the existing sewer. In other circumstances there may be an express requirement to fill these voids as a separate operation in order to control their stabilization more accurately. This situation is covered by classification Y 2 3 *. Two separate items are required, one for the number of holes (which includes either forming and packing around a purpose-made hole or packing around a suitable existing crack) and a second item for the volume of grout injected. Holes are not measured for annulus grouting. Where the Engineer does not expressly require the use of a particular method of grouting the volumes of annulus and void grouting may be combined at classification Y 3 6 0. If there is an express requirement to grout voids separately, classification Y 2 3 * is used. This is the effect of rules M4 and M5.

The cost of locating and reconnecting live laterals to a relined sewer is significant. It can be priced against the items given at classification Y 4 1 *. Sometimes an existing lateral may need to be adjusted to vary its gradient. Such regrading of laterals is covered by rule A10 and, provided that the regrading is limited to the final metre of the lateral, all the associated regrading costs are included in the item (rule C6). Where the regrading extends beyond 1 m, separate items for the work are required. An item for jointing laterals (Y 4 1 *) is also needed where there is a branch in the relined sewer.

Classification Y 6–8 * * covers installed, altered and abandoned manholes and other permanent shafts and chambers. The rules follow closely those given for manholes in class K. Modifications are made to reflect the special characteristics of working around renovated and abandoned sewers. Note that the items at classification Y 6 * * include the cost of removing existing manholes (rule C9) as well as the cost of providing a replacement manhole. Full particulars of work to existing manholes are required in item descriptions (rule A13).

Interruptions are measured at classification Y 8 * *. These items are intended to help Engineers who, having access to sound data about previous sewer flows, wish the Employer to assume the risk of flash floods. The Engineer should express a requirement for the Contractor to provide a minimum pumping capacity which he considers adequate. Interruptions will then only be measured when the sewer flows exceed the installed pumping capacity (rule M6). The installed capacity may, of course, exceed the minimum expressed requirement. The interruption items should cover the costs associated with plant and labour under utilized. They should not cover consequential costs such as the cost of making good a sewer which has been damaged as a result of a flash flood. Rules A1 and A2 subdivide the interruption items by location and sewer type and size.

*Schedule of changes in CESMM2*

Class Y is a complete new class introduced into the CESMM.

| Number | Item description | Unit | Quantity | Rate | Amount £ | p |
|---|---|---|---|---|---|---|
| | NEWTON STREET MANHOLES 25 - 30. | | | | | |
| | EXISTING BRICK SEWER NOMINAL SIZE 600 mm OVOID. | | | | | |
| | Preparation. | | | | | |
| Y110 | Cleaning. | m | 57 | | | |
| | Removing intrusions. | | | | | |
| Y121 | Lateral; bore not exceeding 150 mm; clayware. | nr | 12 | | | |
| Y123 | Dead water main; bore not exceeding 150 mm; cast iron. | nr | 1 | | | |
| | Plugging laterals with grout as Specification clause 7.21. | | | | | |
| Y131 | Bore not exceeding 300 mm. | nr | 21 | | | |
| Y132 | U shaped; internal cross sectional dimensions 350 mm x 200 mm. | nr | 2 | | | |
| | Filling laterals with grout as Specification clause 7.21. | | | | | |
| Y142 | Internal cross section dimensions 450 mm x 350 mm U-shaped. | m3 | 1 | | | |
| | Local internal repairs. | | | | | |
| Y151 | Area not exceeding 0.1 m2. | nr | 6 | | | |
| Y152 | Area 0.1 - 0.25 m2. | nr | 4 | | | |
| Y153 | Area 0.6 m2. | nr | 1 | | | |
| | | | PAGE TOTAL | | | |

NEWTON STREET

| Number | Item description | Unit | Quantity | Rate | Amount £ | p |
|---|---|---|---|---|---|---|
| | EXISTING BRICK SEWER NOMINAL SIZE 600 mm OVOID. | | | | | |
| | Stabilization of existing sewers. | | | | | |
| Y210 | Pointing brickwork with cement mortar. | m2 | 14 | | | |
| Y290 | Pointing pipe joints under pressure with epoxy mortar. | m2 | 7 | | | |
| | External grouting. | | | | | |
| Y231 | Number of holes. | nr | 4 | | | |
| Y232 | Injection of cement grout as Specification clause 7.24. | m3 | 6 | | | |
| | Renovation of existing sewers. | | | | | |
| | Sliplinings. | | | | | |
| Y311 | Butt fusion welded HDPE type SDR 4, thickness 4 mm, 500 mm minimum internal diameter. | m | 57 | | | |
| | | | | PAGE TOTAL | | |

NEWTON STREET

| Number | Item description | Unit | Quantity | Rate | Amount £ | p |
|---|---|---|---|---|---|---|
| | EXISTING BRICK SEWER NOMINAL SIZE 1200 x 900 mm, EGG SHAPED. | | | | | |
| | Renovation of existing sewers. | | | | | |
| | Segmental linings. | | | | | |
| Y333 | Glass reinforced plastic as Specification clause 7.50, minimum internal cross section dimensions 1050 x 750 mm. | m | 45 | | | |
| Y334.1 | Glass reinforced concrete, 15 mm thick, internal cross section dimensions 1050 x 750 mm, egg shaped. | m | 38 | | | |
| Y334.2 | Glass reinforced concrete, 15 mm thick, internal cross section dimensions 1320 x 850 mm, curved to offset of 70 mm per metre. | m | 2 | | | |
| | Annulus grouting. | | | | | |
| Y360 | Cement grout as Specification clause 7.25. | m3 | 10 | | | |
| | Laterals to renovated sewers. | | | | | |
| | Jointing. | | | | | |
| Y411.1 | Bore: not exceeding 150 mm, to HDPE sliplining, type SDR 17. | nr | 10 | | | |
| Y411.2 | Bore: 150 - 300 mm, to HDPE sliplining, type SDR 17, regraded. | nr | 3 | | | |
| Y413 | 900 x 600 mm, egg shaped, to GRP segmental lining, 15 mm thick. | nr | 1 | | | |
| | | | PAGE TOTAL | | | |

NEWTON STREET

| Number | Item description | Unit | Quantity | Rate | Amount £ | p |
|---|---|---|---|---|---|---|
| | EXISTING BRICK SEWER NOMINAL SIZE 1200 x 900 mm, EGG SHAPED. | | | | | |
| | Flap valves. | | | | | |
| Y421.1 | Remove existing; nominal diameter 225 mm. | nr | 2 | | | |
| Y421.2 | Remove existing;nominal diameter 300 mm. | nr | 3 | | | |
| Y422 | Replace existing; nominal diameter 225 mm. | nr | 2 | | | |
| Y423 | New; Holdwater catalogue reference 918; nominal diameter 300 mm. | nr | 3 | | | |
| | New manholes in new locations. | | | | | |
| Y525 | Brick with backdrop type 1A, heavy duty cover and frame; depth 3 - 3.5 m. | nr | 2 | | | |
| Y557 | Precast concrete type 2B, heavy duty cover and frame; depth 7.5 m. | nr | 1 | | | |
| | New manholes replacing existing manholes. | | | | | |
| Y653 | Precast concrete type 2C, heavy duty cover and frame; depth 2 - 2.5 m. | nr | 2 | | | |
| | | | PAGE TOTAL | | | |

NEWTON STREET

| Number | Item description | Unit | Quantity | Rate | Amount £ | p |
|---|---|---|---|---|---|---|
| | EXISTING BRICK SEWER NOMINAL SIZE 1200 x 900 mm, EGG SHAPED. | | | | | |
| | Existing manholes. | | | | | |
| | Abandonment, as Drawing 22/C. | | | | | |
| Y713 | Depth 2 - 2.5 m including removing cover slab, breaking back shaft and backfilling with pulverised fuel ash. | nr | 1 | | | |
| Y717 | Depth 5.5 m including removing cover slab, breaking back shaft and backfilling with pulverised fuel ash. | nr | 1 | | | |
| | Alterations. | | | | | |
| Y720 | Work to benching and inverts, as Drawing 27/G including breaking out, re-haunching and dealing with flows. | nr | 1 | | | |
| | Interruptions. | | | | | |
| Y810 | Preparation of existing sewers. | h | 10 | | | |
| Y820 | Stabilization of existing sewers. | h | 10 | | | |
| | Renovation of existing sewers. | | | | | |
| Y833 | Segmental linings. | h | 20 | | | |
| | | | PAGE TOTAL | | | |

# *Index*